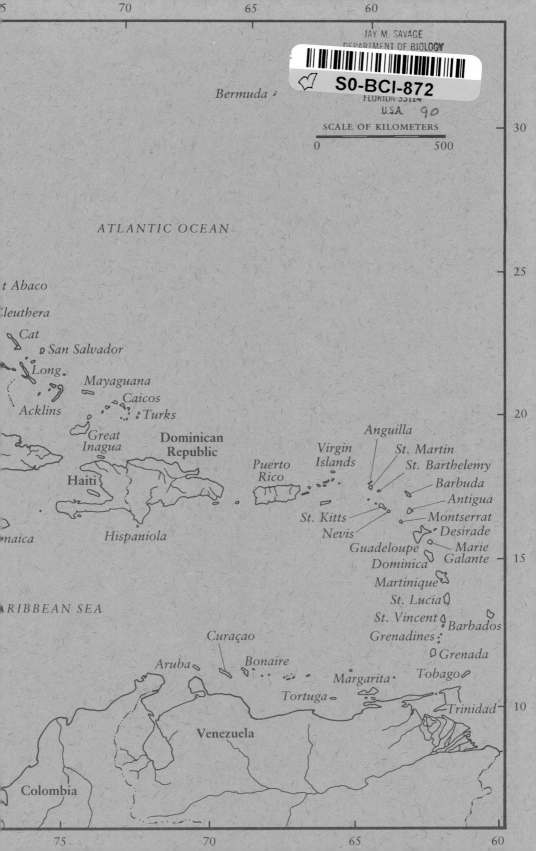

70   65   60

Bermuda

SCALE OF KILOMETERS

0   500   30

ATLANTIC OCEAN

25

t Abaco

leuthera

Cat

San Salvador

Long.

Mayaguana

Caicos

Acklins   Turks   20

Great   Dominican   Anguilla
Inagua   Republic   Virgin   St. Martin
                                   Puerto   Islands   St. Barthelemy
Haiti   Rico   Barbuda
                                                                  Antigua
                         St. Kitts   Montserrat
maica   Nevis   Desirade
        Hispaniola   Guadeloupe   Marie
                                              Dominica   Galante   15
                                             Martinique
RIBBEAN SEA   St. Lucia
                                             St. Vincent   Barbados
                      Curaçao   Grenadines
                                                           Grenada
             Aruba   Bonaire   Tobago
                                      Margarita
                           Tortuga   Trinidad   10

                    Venezuela

Colombia

75   70   65   60

Zoogeography of Caribbean Insects

Zoogeography of Caribbean Insects

EDITED BY

James K. Liebherr

Department of Entomology
Cornell University

Comstock Publishing Associates A DIVISION OF

Cornell University Press   ITHACA AND LONDON

First published 1988 by Cornell University Press.

Library of Congress Cataloging-in-Publication Data

Zoogeography of Caribbean insects.

Includes indexes.
1. Insects—Caribbean Area—Geographical distribution. 2. Insects—West Indies—Geographical distribution. I. Liebherr, James Kenneth.
QL479.A1Z66   1988   595.709729   87-47868
ISBN 0-8014-2143-8 (alk. paper)

Printed in the United States of America
The paper in this book is acid-free and meets the guidelines for permanence and durability of the Committee on Production Guidelines for Book Longevity of the Council on Library Resources.

CONTENTS

PREFACE

The science of biogeography attempts to recount the ecological and historical factors that have interacted to produce biotic diversity. From empirical data on geographic distributions and from ecological and phylogenetic relationships of taxa, the biogeographer hopes to derive specific hypotheses for the histories of taxa of interest as well as more general principles helpful in elucidating the histories of unstudied groups. This book adds to the available information on insect distribution in a well-studied biogeographic region, the Caribbean, and serves as a report on the various methods used to analyze insect distributional data.

The diversity of insects far outclasses that of any other taxon in the Caribbean. This diversity has occupied much of the time of insect systematists, who must complete revisionary systematic research before they can begin to contribute new data to the field of biogeography. But the wealth of insect species and, in many cases, the restricted distributions of insect taxa suggest that biogeographic data from insects will dwarf the data obtainable from other taxa.

This book is a progress report on our understanding of the origins of the Caribbean insect fauna. To date, no other volume on biogeography has concentrated on insects from this region. It is our hope that the presentation of data on a variety of insect groups will illustrate various possibilities for biogeographic research using insects. Biological and distributional attributes differ among the taxa discussed, leading the contributors to use different techniques of analysis for the various groups.

In addition to the entomological contributions in this book are chapters by Thomas Donnelly, who provides a critical review of the geologic literature, and Edward Connor, who critiques the various techniques used by biogeographers. Thus the book aims to broaden the empirical data base on

Caribbean insects and to aid understanding of the context in which these data are interpreted.

Most of the contributions in this volume were presented at a symposium, "Historical Biogeography of the Caribbean Insect Fauna," held at the Entomological Society of America national meetings in Hollywood, Florida, on 10 December 1985. Manuscripts were then prepared and submitted to outside review either by the authors or by me. In submitting our manuscripts to other systematists and biogeographers knowledgeable about the taxa covered and about biogeographic techniques, we hope we have ameliorated flaws common to many symposium volumes.

I thank the Entomological Society of America, for sponsoring the initial symposium, and Carl Schaefer in particular, for championing the symposium topic. Robb Reavill of Cornell University Press has provided encouragement throughout the project. I thank all the contributors and reviewers for adhering to deadlines I often had trouble meeting.

JAMES K. LIEBHERR

Ithaca, New York

CONTRIBUTORS

Edward F. Connor, Department of Environmental Sciences, University of Virginia, Charlottesville, VA 22903

Thomas W. Donnelly, Department of Geological Sciences and Environmental Studies, State University of New York, Binghamton, NY 13901

George C. Eickwort, Department of Entomology, Cornell University, Ithaca, NY 14853-0999

David A. Grimaldi, Department of Entomology, Cornell University, Ithaca, NY 14853-0999; current address Department of Entomology, American Museum of Natural History, Central Park West at 79th St., New York, NY 10024

Steven W. Hamilton, Department of Entomology, Clemson University, Clemson, SC 29634-0365; current address Department of Biology, Austin Peay State University, Clarksville, TN 37044

James K. Liebherr, Department of Entomology, Cornell University, Ithaca, NY 14853-0999

Stephen W. Nichols, Department of Entomology, Cornell University, Ithaca, NY 14853-0999

J. A. Ramos, Department of Biology, University of Puerto Rico, Mayagüez, PR 00709

James A. Slater, Department of Ecology and Evolutionary Biology, University of Connecticut, Storrs, CT 06268

Edward O. Wilson, Museum of Comparative Zoology, Harvard University, Cambridge, MA 02138

Zoogeography of Caribbean Insects

1 · The Caribbean:
Fertile Ground for Zoogeography

James K. Liebherr

The Caribbean is a complex geographic region, with a geologic history as complex as its present-day configuration. The Caribbean Sea is ringed by land: Mesoamerica to the west, South America to the south, the string of Lesser Antilles to the east, and the four Greater Antilles and associated islands to the north. The Greater Antilles and the larger islands of the Lesser Antilles possess substantial mountainous areas that support subtropical wet to montane wet forests depending on elevation. Nonetheless, the Antilles have been considered faunistically depauperate by most zoogeographers working with vertebrate taxa. As Wallace (1876:61) wrote, "There are probably no land areas on the globe, so highly favoured by nature in all the essentials for supporting animal life, and at the same time so poor in all the more highly organized groups of animals."

Because of the relatively limited and easily circumscribed terrestrial vertebrate fauna, the complex geography, and the presence of colonial lands of most of the major Western political powers, the West Indies have served as a focal point for a broad array of zoogeographers. These scientists have brought their own skills and biases to bear on the study of this region. As such, a review of the zoogeographic approaches to study of the Caribbean can serve as a review of the science of zoogeography.

This book serves as an overview of present-day zoogeographic thought about the West Indies, using insects as the study organisms. Over a hundred years ago it was said, "The various West Indian islands have not been well explored entomologically; one reason no doubt being, that their comparative poverty renders them little attractive to the professional collector, while the abounding riches of Central and South America lie so near at hand" (Wallace 1876:73). Even though professional collectors could not make a living off the West Indian insect fauna, we should not conclude that the fauna is

1

depauperate. Many Antillean insect taxa can be characterized as sparse in numbers of individuals but highly endemic and taxonomically diverse. Only as a result of concerted effort by many entomologists can we say today that we have an entomological data base useful for zoogeographic analysis. The high incidence of endemic taxa suggests that continuing field research will be necessary before we can feel comfortable with our knowledge of the geographic distribution and diversity of the West Indian insect fauna.

Review of West Indian Zoogeography

Land Bridges versus Overwater Dispersal

The earliest school of zoogeographic thought claiming to understand the origin of the West Indian fauna sought to explain faunistic similarity among areas by the past occurrence of land bridges. Allen (1911) analyzed the Antillean mammal fauna, and wherever species or genera were held in common in two now-isolated areas, he hypothesized a land-bridge connection. In his analysis, successive land bridges were required to bridge Jamaica and Honduras; Yucatán, Cuba, and Hispaniola; Jamaica, Hispaniola, Puerto Rico, and the Bahamas; and Cuba and Florida. The relative timing of past land connections was set by the degree of taxonomic divergence shown by the inter-area faunal identities. The existence of land bridges was supported by the geologic literature.

Thomas Barbour (1914), in his analysis of the Jamaican herpetofauna, likewise used the geologic conclusions of Hill (1899) to bolster his argument for the past occurrence of land bridges. Barbour (1916) considered the geologic data to be strongly supportive of land bridges, citing the occurrence of manganese nodules and radiolarians at 1200 m elevations in the mountains of Borneo as indications that much of what is now dry land was once under water. Of course, such data are now interpreted in the very different context of plate tectonics. Even when using geologic data to substantiate zoogeographic ideas, Barbour believed the biological data were paramount. Of geologic data he wrote: "To the zoölogist these geologic problems seem so differently interpreted by different and equally gifted and trustworthy students that one is inclined to relegate them all to the limbo of where 'you pay your money and take your choice'" (Barbour 1916:152, in Matthew 1939).

The biological data used to support the past existence of land bridges in the Caribbean included (1) the homogeneous nature of the Antillean fauna; (2) the improbability to impossibility of overwater dispersal of salt-sensitive taxa; (3) the many elements of the fauna, too numerous to credibly support overwater dispersal of all of their ancestors; (4) the inability of the various endemic species to evolve if rafting were rampant; (5) the role of extinction in transforming the originally diverse island faunas into their present-day impoverished condition.

The alternative school of thought to the land-bridge proponents was comprised of those who advocated overwater dispersal of individual taxa as the principal means of building up the Antillean fauna. A pioneer of this school was William D. Matthew (Simpson 1956), who presented his interpretation of the origin of Antillean mammals in 1918. His criteria for reasoning that overwater dispersal better explained the colonization of the Antilles included (1) the limited number of land-mammal groups occurring on the Antilles (insectivores, rodents, and edentates); (2) the different times of origin of these groups on the Antilles, based on levels of taxonomic divergence and fossil evidence; (3) the lack of any geologic evidence supporting connections of the islands to the mainland. It is of interest that Matthew conceded to a past union of the Greater Antilles based on the faunal similarities of those islands.

Philip Darlington (1938:274) expanded Matthew's arguments for overwater colonization of the Antilles, developing his "Theory and Mathematics of Dispersal of Organisms across Water Gaps." His central thesis was that dispersal was not an accidental or random process but could be inferred through knowledge of the width of water gaps, island areas and ages, directions of storms and currents, and the biological attributes of potential colonizers. He opposed the idea that dispersed taxa show erratic distributions, whereas land connections produce orderly patterns. He argued that islands would be colonized in a stepwise fashion, leading to an ordered fauna. Thus erratic dispersal patterns would be due as much to differing suitability of various islands to continued habitation as to rampant dispersal. In all of these arguments, Darlington presaged the theory of island biogeography of MacArthur and Wilson (1967).

Darlington countered the arguments against overwater dispersal by presenting patterns of diversity of a variety of taxa. Dispersal was considered to start at areas of high taxonomic diversity and to be most recent in areas of low diversity. Thus the length of time a taxon inhabited an area was determinable by its diversity there. Dispersal routes were plotted for a variety of groups, with major routes determined by the majority of the groups. The land masses were considered stable, requiring taxa to move from island to island. By charting paths of biotic attenuation, Darlington (1957) suggested that the Antillean fauna is derivable from three dispersal routes (Fig. 1-1): (1) Yucatán to Cuba; (2) Central America to Jamaica; (3) South America to the Lesser Antilles.

Darlington (1938) also used distributional formulas to decide between colonization by means of a land-bridge corridor or by overwater dispersal. Islands were connected by a network of dashes, with long dashes indicating low faunal similarity. This species-similarity formula was then compared to

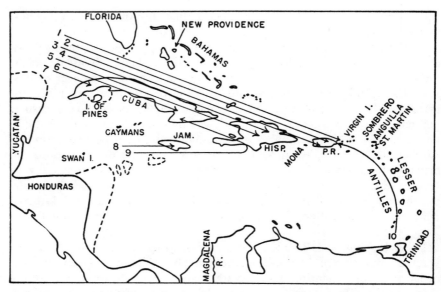

Figure 1-1. The West Indies. Broken lines show banks off Central America that may have been land in the past. Numbered arrows show distance and (apparent) direction of dispersal on the Greater Antilles of the following animals: 1, *Bufo* (toads); 2, ground sloths; 3, *Nesophontes* (small insectivores); 4, cichlid fishes; 5, *Solenodon* (larger insectivores); 6, atelopodid frogs; 7, gar pikes; 8, monkeys; 9, *Hyla* (frogs); 10, *Leptodactylus* (frogs). This list includes all secondary-division fishes of the Greater Antilles except Cyprinodontes, all amphibians except *Eleutherodactylus*, and all flightless land mammals except rodents. The arrows therefore illustrate the limited diversity of the Greater Antillean fauna, its orderliness, and the fact that the most significant parts of it form a simple pattern of apparent immigration, mostly from Central America, with Cuba the most important port of entry (from Philip J. Darlington Jr., *Zoogeography*, Fig. 63, copyright © 1957 by John Wiley & Sons, Inc., reprinted by permission of John Wiley & Sons, Inc.).

a geographically based formula, in which the sizes of watergaps were represented by dashes of various lengths. Darlington concluded that Jamaica's fauna is more similar to Hispaniola's than to Cuba's, even though Jamaica is closer to Cuba than to Hispaniola. He suggested that Jamaica, Hispaniola, and Cuba may have been connected by land or separated by only narrow water gaps, citing high faunal similarities supporting this formula by use of data from a variety of taxa. Thus, as in Matthew (1918), interisland vicariance was admitted as probable. By 1957, however, Darlington had abandoned any belief in vicariance in the Caribbean, believing that geologic and biological data should be interpreted separately. With his methods the biological data suggested dispersal as the organizing force in the Antillean fauna.

George Gaylord Simpson (1956) produced the final synthesis of arguments supporting overwater dispersal instead of past land-bridge connections. He had a more extensive data base available to him than did Matthew (1918); but using similar methods, he came to similar conclusions. Based on his data, only seven or eight ancestors need have colonized the Greater Antilles to produce the known land-mammal fauna, both fossil and present-day. Like Matthew, Simpson regarded fossils of Antillean groups or their mainland relatives as evidence for the time of colonization of the Antilles. Thus the insectivores *Nesophontes* and *Solenodon* could be derived from North American relatives from the late Cretaceous onward. The edentates could be derived from Miocene South American groups. These megalonychid sloths entered North America from South America prior to the late Pliocene closure of the Panamanian isthmus, suggesting that they rafted to Cuba from Colombia. The Antillean rodent genera are derivable from South American groups first seen in the Miocene. The only fossil primate, *Xenothrix*, is known from Jamaica and is of Central or South American affinities.

Additional data determining relative time of colonization included information on whether the Antillean taxa are endemic at the family or the generic level. Linear filtering dispersal was generally assumed, based on the attenuation of families and genera leading from Cuba or Jamaica to Hispaniola and to Puerto Rico.

Plate Tectonics and Vicariance Biogeography

The development of plate tectonic theory provided the mechanism for what had been hypothesized years before by Wegener (1915)—that the continental land masses are mobile, not stable. Such a finding was resisted by dispersalist biogeographers or, if accepted, said to be irrelevant because dispersal subsequent to plate motion produced present-day patterns (Darlington 1957). Plate tectonic reconstructions of the Antillean region (e.g., Malfait and Dinkelman 1972) were used by Rosen (1975) as the geologic basis for a vicariance model of Caribbean biogeography. Rosen used the biogeographic methods of Croizat et al. (1974), in which distributions of monophyletic groups are circumscribed on present-day geography. The resultant connected distributions are called tracks. Where a number of tracks are coincident, or a number of partially overlapping individual tracks agree in part, they are said to compose a generalized track. Where tracks connect currently disjunct distributions, a geologically based hypothesis for the disjunction, or vicariance, is proposed. In this manner, timing of vicariance is determined strictly by the geologic data. Taxonomic diversity is discounted, and diversity becomes the result of unknowable evolutionary rates. Fossils are thought useful only to provide minimum ages of taxa and to serve as evidence of past distributions.

Generalized tracks are nothing more than undirected connections of the ranges of taxa within a monophyletic group and are therefore based on a certain level of phyletic similarity. Testing the congruence of such tracks without knowledge of the cladistic relationships within the monophyletic groups is little different from what was done by land-bridge or dispersal advocates (McDowall 1978). But the vicariance biogeographic approach added a new technique to the analysis—the ordering of relationships within each monophyletic group using cladistic criteria, that is, grouping taxa by their shared-derived character states, also called synapomorphies (Hennig 1966). These directed, branching networks can be compared, using component analysis, to determine whether the branching sequences are congruent (Nelson 1979). By comparing the cladistic relationships of taxa composing the various individual tracks, Rosen (1975) introduced a totally biological means of substantiating vicariance hypotheses. If enough cladograms of different groups occupying the same areas prove congruent, the pattern of area relationships is probably due to a common cause—a common sequence of vicariance events. Rosen assumed dispersal of taxa to produce random distributions, implying that vicariance and dispersal hypotheses would be easily distinguishable.

In 1975 Rosen did not compare cladograms of different groups to test his vicariance model. Instead, he proposed a geologic hypothesis for the region and tested the animal relationships against it. Donnelly (see chap. 2) reviews the various aspects of recent geologic models for the origins of the Caribbean. Rosen's (1975) geologic model approximates what Donnelly calls the "most" mobilist model.

Rosen (1975) recognized four general tracks plus one minor track for the Caribbean. The general tracks are a South American–Caribbean track, a North American–Caribbean track, an East Pacific–Caribbean track, and an Eastern Atlantic or West African–Caribbean track. The fifth minor track is a dispersal track with a source in South America, and dispersal to the Lesser and Greater Antilles.

The various tracks could be identified by the cladistic relationships of the Antillean and mainland taxa. The South American–Caribbean track taxa would exhibit a cladistic position basal to a South plus Central American monophyletic group (Fig. 1-2A). Antillean groups would have colonized a proto-Antillean archipelago between Central and South America some time in the early Cenozoic and vicariated via eastward tectonic movement of the Antilles. Central America was subsequently reinvaded by more South American taxa via range expansion during and after the closing of the Isthmus of Panama. The North American–Caribbean track (Fig. 1-2B) is comprised of taxa in which Antillean members are the sister groups to taxa in North, northern Central, and Central America, northern Central America being the Chiapan to Honduran highlands, and Central America ranging from Costa Rica through Panama. In this case Antillean taxa would have vicariated as the Antilles broke off of southern Yucatán or Central America.

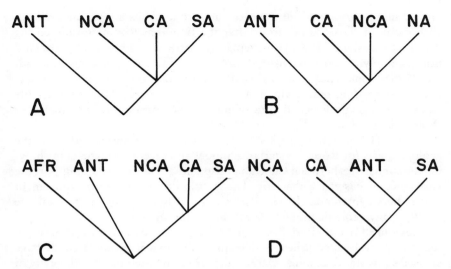

Figure 1-2. Area cladograms representing relationships expected for taxa in various Antillean general tracks. (A) Vicariance-based relationships of South American–Caribbean track; (B) Vicariance-based relationships of North American–Caribbean track; (C) Old Southern groups showing African relationships; (D) Dispersal track with Antillean taxa derived overwater from South America. ANT, Antilles; NCA, northern Central America; CA, Central America; SA, South America; NA, North America; AFR, Africa.

The Eastern Pacific–Caribbean track deals principally with marine organisms separated by the Isthmus of Panama. It also includes terrestrial taxa with Galapagos-Antillean relationships. The West African–Caribbean track includes taxa with representatives in the Gulf of Guinea area. I refer to these as Old Southern groups (Fig. 1-2C) and consider this track to be a variation of the South American–Caribbean track, but with taxa of sufficient age to preserve a record of Cretaceous Amphi-Atlantic vicariance. Such an interpretation is useful for terrestrial insect groups; because Rosen was principally concerned with Atlantic oceanic taxa, he did not include North or South America in this track.

Dispersal onto the oceanic island chain of the Lesser Antilles would produce the relationships shown in Figure 1-2D: Antillean groups most closely related to South American taxa, and Central American and northern Central American taxa less closely related.

Whereas previously published major expositions on the Antillean fauna appeared at approximately 20-year intervals with little change in opinions over nearly half a century (Matthew 1918, Darlington 1938, Simpson 1956), Rosen (1975) stimulated extensive discussion within the past 10 years. Pregill (1981) criticized Rosen's model on its geologic basis and by citing fossil evidence showing that the Antillean vertebrate fauna is comprised of a limited number of taxa that appeared at various times in the fossil record, with many of those dating from after the time that Rosen considered the Antilles to be islands. Pregill thus replayed Matthew's (1918) critique of land bridges, applying more extensive data to the vicariance arguments. MacFadden (1981) took exception to Pregill's interpretation of the data on insectivores, suggesting that a vicariance hypothesis best explained insectivore data, even though dispersal data might also be appropriate for other taxa (MacFadden 1980).

Hedges (1982) cited various geologic papers that supported both the geologic basis for Rosen's vicariance model and recent geologic data showing the similarities between Cuba and Hispaniola. The latter data support the vicariance of eastern Cuba from Hispaniola due to transform movement along the Cayman trench. He also cautioned biogeographers to evaluate carefully any geologic model before using it for analysis.

Briggs (1984) criticized Rosen for basing his vicariance hypothesis on secondary freshwater fishes, which on physiological grounds could have crossed saltwater to colonize the Antilles. He also noted that the diversity of endemic genera and species attenuated from Cuba eastward in the Greater Antilles. He reiterated the dispersal school's contention that different taxonomic levels of endemism imply different times of origin for the Antillean taxa, thus summarizing Darlington's (1957) major points supporting dispersal.

Recently, several modified vicariance hypotheses for the Caribbean have been presented. Savage (1982) analyzed the Central American and Antillean herpetofauna, concluding that the Antilles provided a connection between North and South America early in the Cenozoic. It was considered unimportant whether this connection was an island chain or continuous land, as it provided the means for concordant dispersal of numerous South American taxa to Central America. Savage's model emphasized the nonrandom, concordant nature of dispersal given a linear archipelago or a terrestrial corridor. He concluded that the data from angiosperms, fishes, amphibians, and reptiles are concordant. Mammals do not exhibit the same patterns, because only marsupials were extant in South America at the time of the Cenozoic vicariance of North and South America and the Antilles.

Rosen (1985) proposed a revised and more explicit hypothesis of vicariance relationships within the Antillean fauna, based on several recent mobilist geologic models (e.g., Sykes et al. 1982). The "most" mobilist models of Caribbean geology (Donnelly, chap. 2) set the Eocene as the period of formation of several of the Antilles from portions of Central America. They also support interisland vicariance and intraisland hybridization of Cuba and Hispaniola. Liebherr (see chap. 6) summarizes these data and their implications for some carabid beetle distributions. Hamilton (see chap. 7) notes similar taxon-area relationships in *Polycentropus* caddisflies.

Current Points of Contention

It is clear that if we are to make advances in understanding Caribbean biogeography, we must refine our techniques of analysis and obtain more data from previously unused taxa. Biogeographic analysis has advanced intermittently since the 19th century as we learned more about the biological attributes of organisms and how they interact with the environment, and as we learned more of the phylogenetic affinities of organisms and how best to quantify those affinities. The utility of cladistic analysis as a basis for biogeographic studies is now well documented (Nelson and Platnick 1981), owing to the logical consistency of testing congruent cladograms of taxa and areas using component analysis (Nelson 1979, Humphries 1982, Humphries and Parenti 1986). The alternative of grouping areas using pairwise indexes of similarity does not allow such a test. The assumptions used to explain incongruence and the significance of various levels of congruence are other topics that require attention (Simberloff et al. 1981, see also Connor, chap. 11). The use of statistical methods for determining significant patterns of distribution can be considered a subset of the larger question of whether statistical significance can be assigned to historical phenomena such as phylogenies. One might adopt the stance that the most parsimonious solution is correct, obviating the need for tests of significance. Alternatively, one may recognize that in all phylogenetic analyses, choice of taxa and characters may bias the outcome. Determining whether the observed patterns differ from some predetermined null hypothesis provides a means to test the validity of the interpretation of results.

Various other biogeographic assumptions need to be examined carefully if we are to produce meaningful studies. What is the nature and extent of dispersal? Are its patterns random, or are they patterned on geography past and present? Population-level studies within wide-ranging taxa may provide the resolution needed to trace gene flow (Liebherr 1985, see also Slater, chap. 3), showing that widespread taxa are informative, at least with respect to dispersal. What are the roles of fossils in biogeographic analysis (see Wilson, chap. 9)? Under what circumstances (if any) can taxonomic diversity and levels of taxonomic divergence be used in biogeographic analysis?

If biogeographic analysis is firmly wedded to geologic data, how can we choose among the possibilities, except to do as Thomas Barbour (1916) proposed? Rosen (1985) and Donnelly (chap. 2) proposed to find points of consensus among a variety of models and, in Donnelly's case, to reduce the confusion to a comprehensible number of alternatives. Also, if we are to support the independence of biogeographic and geologic data, how can we analyze biogeographic data to suggest new geologic alternatives (see Connor, chap. 11)?

The sedimentary record overlies the tectonic history of the Caribbean.

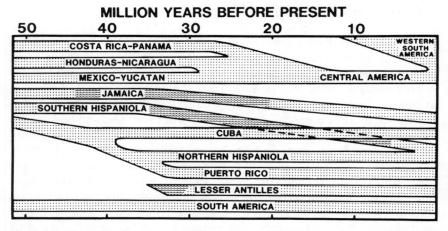

Figure 1-3. Hypothetical land-mass relationships in the Caribbean during the last 50 million years based on geologic models discussed in Buskirk (1985). The time-line sketch indicates relative positions of the labeled areas. Increased distance apart indicates barriers to dispersal, and convergences indicate reduced dispersal barriers. Areas that were largely inundated are designated with water-wave symbols (Fig. 4 of Buskirk 1985, reproduced by permission of Blackwell Scientific Publications Limited).

Movements of land masses may be accounted for by slippage along faults and by earthquake data, but does the sedimentary record support the continuous exposure of various land forms, making them habitable by terrestrial organisms? Buskirk (1985) presented data showing that even though Jamaica and southern Hispaniola may be the result of old land forms, their history includes substantial periods under water (Fig. 1-3). As noted by Hedges (1982), such data may be cause for caution in constructing vicariance hypotheses. Nonetheless, geologic data often cannot provide unequivocal answers about the history of areas, making biological data of utmost importance in the interpretation of faunal histories.

Pleistocene climatic changes have been used by dispersal biogeographers to explain many of the taxonomic radiations in the Antilles. The vicariance approach does not exclude Pleistocene speciation, and so we can expect future biogeographic work to elucidate patterns related to Pleistocene climatic oscillations. Hispaniola, with its three parallel cordilleras and probable hybrid history, should prove of special interest in this regard. Munroe (1950) provided an early vicariance study of a West Indian insect taxon, the wood satyr butterfly genus *Calisto* (Lepidoptera: Satyridae). *Calisto* species are generally restricted to portions of Hispaniola, with Pleistocene speciation highly probable. The vicariance approach frees biogeographers from insisting on Pleistocene causes of speciation if speciation patterns do not fit. As more studies are completed, a more continuous distribution of times of speciation from the Mesozoic through the more recent periods will probably result.

Insects as a Data Source

The Antillean insect fauna offers many advantages for biogeographic analysis not present in the vertebrate fauna. Many modern insect orders and families were in existence by the Jurassic, and nearly all orders were extant by the Cretaceous (Riek 1970). Insects are especially likely to help clarify old patterns of faunal relationships.

The Old Southern groups (Fig. 1-2C), in which New World taxa have Old World affinities, are well represented in the Antillean insect fauna. Added to the list of Antillean taxa with Old World affinities, which included the dam-selfly *Phylolestes* (Odonata) (Flint 1978), the nymphalid butterfly *Archimes-tra* (Lepidoptera) (Munroe 1949), the caddisfly *Antillopsyche* (Trichoptera) (Flint 1978), and the ground beetle *Barylaus* (Coleoptera) (Liebherr 1986), are the lygaeid bug *Pachygrontha* (Hemiptera) (see Slater, chap. 3), the ground beetle *Antilliscaris* (Coleoptera) (see Nichols, chap. 5), and the fruit fly *Mayagueza* (Diptera) (see Grimaldi, chap. 8). As more entomologists see these patterns, more Old World Caribbean relationships will be recognized.

The question of whether dispersal or vicariance contributes most to the diversification of the Antillean fauna will remain a subject of controversy. Vicariant relationships among the Greater Antilles are suggested by Hamilton (see chap. 7). Often we know so little about the mainland fauna that substantiation of cladistic relationships expected of Antillean vicariant groups (Figs. 1-2A, B) must await further studies. In some groups, Greater Antillean taxa are more closely related—sometimes conspecific—with South or Central American taxa (e.g., Fig. 1-2D). Such relationships are best explained via overwater dispersal of taxa onto the islands. Eickwort (see chap. 10) presents data on sweat bees (Hymenoptera), and Slater (see chap. 3) presents data for lygaeid bugs (Hemiptera) that conform to patterns of overwater dispersal.

The proliferation of endemic species on islands is best illustrated with insect taxa. Existence on the limited areas of islands can lead to loss of vagility, increased endemicity, and radiation of species. Autochthonous radiations often result in smaller islands hosting larger numbers of species for various taxa. Thus, for the leafhopper subfamily Cicadellinae (Homoptera), Puerto Rico hosts more single-island endemics than do any of the larger Antilles (see Ramos, chap. 4). As an example of a group with less-developed autochthonous radiations, Wilson (see chap. 9) illustrates the correlation of colonization and extinction in Hispaniolan ants. Even here, where Antillean ant-species diversity conforms to that predicted from island area, 40% of the Antillean species are single-island endemics.

With such staggering levels of endemicity, entomologists have an extensive potential data base for intraisland analyses. More extensive field work is required to permit characterization of species distributions and habitat requirements. Even though we have so many insular endemics that are useful for biogeographic analysis, destruction of habitat makes it progressively harder to study known endemics and may eliminate untold numbers of taxa before they are ever recognized. Just as sensitive aquatic species have been used as indicators of environmental degradation in developed countries, distributions of endemic species can show geographic regions that must be preserved to maximize preservation of the Antillean biota. Whereas historical biogeographers are concerned with the forces that have acted to produce our present-day biota, we are obliged to use our skills to protect as much of that biota as we can.

Acknowledgments

I thank all collaborators in this book project for meeting deadlines and for subjecting their manuscripts to outside review. Ed Connor, Mike Bowers, and the students and staff of the University of Virginia Blandy Experimental Farm proved extremely hospitable during the writing of this chapter. I thank Ed, Norman Platnick, and Quentin Wheeler for constructive criticism, and Susan Pohl and Velvet Saunders for their word-processing acumen. Preparation of this manuscript was supported by Hatch Project NY (C) 139406.

References

Allen, G. 1911. Mammals of the West Indies. Bull. Mus. Comp. Zool. 54:175–263.

Barbour, T. 1914. A contribution to the zoogeography of the West Indies, with especial reference to amphibians and reptiles. Mem. Mus. Comp. Zool. 44:209–359 + 1 pl.

———. 1916. Some remarks upon Matthew's *Climate and Evolution*, with supplementary note by W.D. Matthew. Ann. New York Acad. Sci. 27:1–15 (reprinted in Matthew 1939).

Briggs, J.C. 1984. Freshwater fishes and biogeography of Central America and the Antilles. Syst. Zool. 33:428–435.

Buskirk, R.E. 1985. Zoogeographic patterns and tectonic history of Jamaica and the northern Caribbean. J. Biogeogr. 12:445–461.

Croizat, L., G. Nelson, and D.E. Rosen. 1974. Centers of origin and related concepts. Syst. Zool. 23:265–287.

Darlington, P.J., Jr. 1938. The origin of the fauna of the Greater Antilles, with discussion of dispersal of animals over water and through the air. Quart. Rev. Biol. 13:274–300.

———. 1957. Zoogeography: The Geographical Distribution of Animals. John Wiley and Sons, New York. xi + 675 pp.

Flint, O.S., Jr. 1978. Probable origins of the West Indian Trichoptera and Odonata faunas. Proceedings of the 2d International Symposium on Trichoptera, pp. 215–223. Dr. W. Junk, The Hague.

Hedges, S.B. 1982. Caribbean biogeography: Implications of recent plate tectonic studies. Syst. Zool. 31:518–522.

Hennig, W. 1966. Phylogenetic Systematics. University of Illinois Press, Urbana. xi + 263 pp.

Hill, R.T. 1899. The geology and physical geography of Jamaica: Study of a type of Antillean development. Bull. Mus. Comp. Zool. 34:1–226 + 41 pls.

Humphries, C.J. 1982. Vicariance biogeography in Mesoamerica. Ann. Missouri Bot. Gard. 69:444–463.

Humphries, C.J., and L.R. Parenti. 1986. Cladistic Biogeography. Clarendon Press, Oxford. 98 pp.

Liebherr, J.K. 1985. Univariate analysis of multivariate data to evaluate circular overlap in the *Agonum decorum* complex (Coleoptera: Carabidae). Ann. Entomol. Soc. Amer. 78:790–798.

———. 1986. *Barylaus*, new genus (Coleoptera: Carabidae) endemic to the West Indies with Old World affinities. J. New York Entomol. Soc. 95:83–97.

MacArthur, R.H., and E.O. Wilson. 1967. The Theory of Island Biogeography. Monog. Pop. Biol. No. 1. Princeton University Press, Princeton, N.J. xi + 203 pp.

McDowall, R.M. 1978. Generalized tracks and dispersal in biogeography. Syst. Zool. 27:88–104.

MacFadden, B.J. 1980. Rafting mammals or drifting islands? Biogeography of the Greater Antillean insectivores *Nesophontes* and *Solenodon*. J. Biogeogr. 7:11–22.

———. 1981. Comments on Pregill's appraisal of historical biogeography of Caribbean vertebrates: Vicariance, dispersal, or both? Syst. Zool. 30:370–372.

Malfait, B.T., and M.G. Dinkelman. 1972. Circum-Caribbean tectonic and igneous activity and the evolution of the Caribbean plate. Geol. Soc. Amer. Bull. 83:251–272.

Matthew, W.D. 1918. Affinities and origin of the Antillean mammals. Bull. Geol. Soc. Amer. 29:657–666.

———. 1939. Climate and Evolution. Special Publ. New York Acad. Sci., vol. 1. xii + 223 pp.

Munroe, E. 1949. A new genus of Nymphalidae and its affinities (Lepidoptera, Rhopalocera). J. New York Entomol. Soc. 57:67–78.

———. 1950. The systematics of *Calisto* (Lepidoptera, Satyrinae), with remarks on the evolutionary and zoogeographic significance of the genus. J. New York Entomol. Soc. 58:211–240.

Nelson, G. 1979. Cladistic analysis and synthesis: Principles and definitions, with a historical note on Adanson's *Familles des Plantes* (1763–1764). Syst. Zool. 28:1–39.

Nelson, G., and N.I. Platnick. 1981. Systematics and Biogeography: Cladistics and Vicariance. Columbia University Press, New York. 567 pp.

Pregill, G.K. 1981. An appraisal of the vicariance hypothesis of Caribbean biogeography and its application to West Indian terrestrial vertebrates. Syst. Zool. 30:147–155.

Riek, E.F. 1970. Fossil history, *In* C.S.I.R.O. The Insects of Australia, pp. 168–186. Melbourne University Press, Carlton, Victoria.

Rosen, D.E. 1975. A vicariance model of Caribbean biogeography. Syst. Zool. 24:431–464.

———. 1985. Geological hierarchies and biogeographic congruence in the Caribbean. Ann. Missouri Bot. Gard. 72:636–659.

Savage, J.M. 1982. The enigma of the Central American herpetofauna: Dispersals or vicariance? Ann. Missouri Bot. Gard. 69:464–547.

Simberloff, D., K.L. Heck, E.D. McCoy, and E.F. Connor. 1981. There have been no statistical tests of cladistic biogeographical hypotheses, In G. Nelson and D.E. Rosen, eds., Vicariance Biogeography: A Critique, pp. 40–63. Columbia University Press, New York.

Simpson, G.G. 1956. Zoogeography of West Indian land mammals. Amer. Mus. Novit., no. 1759. 28 pp.

Sykes, L.R., W.R. McCann, and A.L. Kafka. 1982. Motion of Caribbean plate during last 7 million years and implications for earlier Cenozoic movements. J. Geophys. Res. 87 (B13):10656–10676.

Wallace, A.R. 1876. The Geographical Distribution of Animals, vol. II. Harper and Bros., New York. 607 pp.

Wegener, A. 1915. Die Entstehung der Kontinente und Ozeane. Brunswig, FRG, Sammlung Vieweg, no. 23. 94 pp.

2 · Geologic Constraints on Caribbean Biogeography

Thomas W. Donnelly

The critical position of Middle America for New World biogeography has led to a century and a half of speculation on the relationship between geologic activity and the distribution of organisms. Charles Darwin, while on the voyage of the Beagle in the 1830s, pondered the role of geologic elevation and subsidence in this area in explaining the history of mammalian faunas in North and South America. More recently this area has fascinated and perplexed generations of biogeographers addressing two basic questions: (1) Did certain organisms originate in North or South America, and when did they migrate to the other continent? and (2) In what manner and at what time did the West Indian islands become populated? For the first question the evidence of vertebrate, especially mammalian, faunas is the best. As summarized in a recent book (Stehli and Webb 1985), there is limited evidence for connections between North and South America in the late Cretaceous, and for separation during the early Cenozoic; a limited Oligocene connection; and an increasing connection from the late Miocene onward, culminating in an extensive faunal exchange in the Pliocene. The question of faunal history on the Antillean islands is more problematical and cannot be addressed directly with fossil evidence, which is limited to a few relatively young occurrences.

The role of the geologist in this controversy is to develop and compare models for the geologic history of Middle America, and to evaluate biological hypotheses according to geologic constraints. Current geologic thinking reflects the profound revolution that swept the profession in the mid-1960s. Before this time, stabilist thinking dominated the profession, with tectonic evolution largely limited to local vertical movements, including the uplift of oceanic areas into emergent zones. Subsequent to the plate-tectonic revolution, a mobilist philosophy has prevailed. However, the Caribbean has remained one of the most controversial areas in the world for geologic reconstructions, because neither rocks nor identifiable geophysical features of sufficiently early ages have been found.

The earlier stabilist views were presented by Woodring (1954), Weyl (1966), and Meyerhoff (1967, 1972). Later papers in this category recognized the importance of plate tectonics but developed Caribbean models largely without plate-tectonic considerations. These include MacGillavry (1970), Weyl (1973, 1980), and Barr (1974). The mobilist views are diverse, differing mainly on the controversial en masse eastward movement of the Caribbean plate from the Pacific Ocean. In chronological order, representative plate-tectonic reconstructions have been presented by Freeland and Dietz (1971), Malfait and Dinkleman (1972), Ladd (1976), MacDonald (1976), Perfit and Heezen (1978), Dickinson and Coney (1980), Tarling (1980), Sykes et al. (1982), Pindell and Dewey (1982), Anderson and Schmidt (1983), Wadge and Burke (1983), Mattson (1984), Duncan and Hargraves (1984), Donnelly (1985), and Pindell (1985). Reference to these is instructive but may draw the reader to the unrealistic conclusion that the major features of geologic history of this area are well understood.

Biogeographers have been quick to utilize plate-tectonics in their analyses. Woodring (1965) realized the consequences of the rise of the Panamanian isthmus for molluscan biogeography. Rosen (1975) provided a seminal reference that stimulated several additional papers (MacFadden 1981, Pregill 1981). Coney (1982) added a geologist's prespective to biogeographic problems. Buskirk (1985) and Durham (1985) have developed new mobilist views into hypotheses centering on Jamaica and the southwestern Caribbean, respectively. All of these works took a dominantly vicariant viewpoint of faunal history.

Throughout this chapter I will summarize inferences developed in Donnelly (1985), where additional bibliographic references can also be found. The most valuable geologic source book is Nairn and Stehli (1975). An important additional source is the Geological Society of America's volume on the Caribbean (see Donnelly et al. 1988). Although not specifically referenced here, that book has profoundly influenced this chapter.

Geologic History of the Caribbean: Areas of General Agreement

Continental Crustal Blocks

This account of geologic and tectonic history is intended more as a guide to a continuing controversy than as a definitive statement. It begins with a summary of areas of general agreement among the various factions.

The identification of ancient continental crust is straightforward and sig-

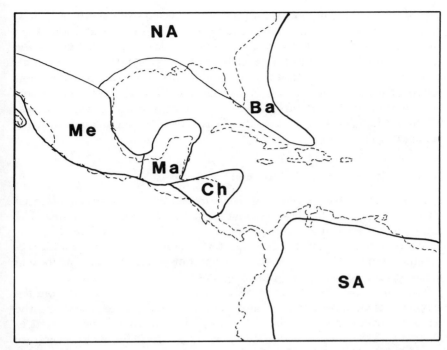

Figure 2-1. Map of the Caribbean region, showing, by continuous line, distribution of pre-Mesozoic continental crust. The three Middle American blocks are the Mexican (Me), Maya (Ma), and Chortis (Ch). Also shown are North America (NA), South America (SA), and the Bahama platform (Ba).

nificant for biogeography because such crust—even in fragments—represents a permanently emergent area. In Middle America, continental crust is confined to the Mexican and Mayan blocks (extending south to central Guatemala) and the Chortis block, consisting of southern Guatemala, Honduras, and most of Nicaragua (Fig. 2-1). Although these crustal areas may have been and undoubtedly were tectonically rearranged during the opening of the Caribbean, they cannot have been below sea level for significant periods of time, except in marginal zones. A possible exception to this generalization is the Nicaraguan Rise portion of the Chortis block, which is discussed below.

Recognition of Oceanic Areas

Just as the recognition of continental crust carries with it the identification of areas of persistent emergence, oceanic crust signifies mainly areas that are persistently submerged. Exceptions are areas that, in thin, curvilinear belts, may, through accumulation of sufficient volcanic and sedimentary materials, achieve emergence at some late stage in their development and are called island arcs. In the Caribbean area, such arcs include Costa Rica, Panama, the Greater Antilles (with the trivial exception of slivers of northernmost Cuba), and the Lesser Antilles. Another poorly understood type of terrane is comprised of slivers of oceanic materials that may be stacked by compressive deformation into linear masses of sufficient thickness to achieve emergence. Much of Cuba and some of the offshore islands of northern South America belong in this category.

Mesozoic Separation of North and South America

There is general agreement that the entire Middle America region formed during a tensional episode when South and North America separated. The age of initial separation is unknown but probably slightly later than for North America and Europe. The oldest Middle American stratigraphic section that seems to be clearly related to separation is early Jurassic and located in east-central Mexico (Schmidt-Effing 1980).

The azimuth and velocities of plate motion have been inferred from the geometry of oceanic magnetic anomalies, oceanic fracture zones, and the relative movement of hot-spots. Seven of these analyses, summarized in Donnelly (1985), agree that for the earliest periods for which there are relevant data (about 165 Ma), South and North America separated between 5 and 6 cm/yr, with South America moving relatively ESE (Fig. 2-2). Separation continued to mid-early Cretaceous (about 125 Ma), when the velocity slowed abruptly as Africa separated from South America, and the South Atlantic Ocean opened. The consequences of inter-American separation were the formation of new oceanic crust and the removal of a terrestrial connection between the Americas.

Concomitant with this intercontinental spreading was extensive deformation of Mexico (with large fragments sliding southeastward relative to blocks to the southwest), the opening of the Gulf of Mexico, and a still controversial reorientation of the Maya block (Mexico east of Tehuantepec, northern Guatemala, Belize) with respect to the remainder of Mexico (Anderson and Schmidt 1983, Pindell 1985).

Geologic History of the Caribbean: Unsettled Controversies of Later Cretaceous and Cenozoic History

The Role of the Caribbean Plate

Perhaps the most controversial of Caribbean tectonic problems is the history of the so-called Caribbean plate, which extends from central Guate-

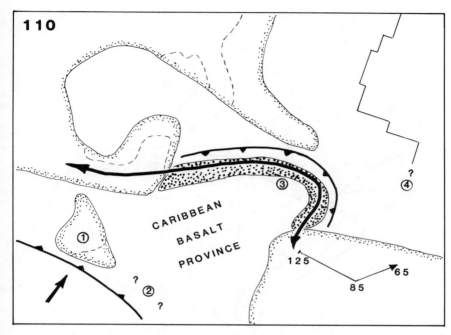

Figure 2-2. Sketch showing a possible configuration of continental fragments and a putative proto-Antillean island arc (heavy stippling) at Albian time, about 100 Ma. The long, double-arrowed heavy line shows a potential connection for terrestrial fauna via a growing island arc formed marginal to the Caribbean basalt province. The shorter arrow indicates movement of the Farallon plate. The broken vector in the lower right shows amount and direction of movement of South America relative to North America between 125 and 65 Ma., with a change in direction at 85 Ma. The heavy lines with triangular barbs indicate subduction zones. Numbered points are: (1) Chortis block, which is a detached continental fragment at this time; (2) the Central American arc, for which no remnants of this age have been identified; (3) the proto-Antillean island arc; and (4) the mid-Atlantic ridge, whose further direction to the southwest or southeast is uncertain at this time.

mala south to Colombia and eastward to the Lesser Antilles. The Greater Antilles and northern South America are highly faulted boundary zones between this plate and North and South America, respectively. Many—probably most—contemporary geologists consider themselves members of the "more" mobilist school and hold that the Caribbean plate moved into its present position from the eastern Pacific, carrying at its front margin the Greater Antilles, and along its rear margin southern Central America. In support of this idea is the inference that a third plate—the now nearly vanished plate of the eastern Pacific called the Farallon plate before about 25 Ma, and the Nazca and Cocos plates after that time—appears to have moved northeastward at much higher velocities than were experienced by either of the American plates (Duncan and Hargraves 1984). The inferred high ve-

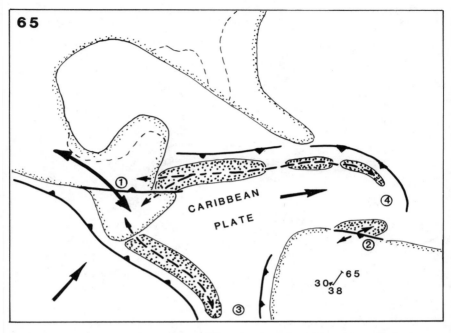

Figure 2-3. Sketch showing a possible configuration of continental fragments and island arc at the end of the Cretaceous. Symbols as in Figure 2-2, except that broken, double-arrowed lines are possible faunal connections that extend on to island-arc segments. Eastward movement of the Caribbean plate has stretched and broken the formerly continuous proto-Antillean island arc into several discrete segments. Numbered points are (1) late Cretaceous suture zone between the Chortis and Maya blocks, which allows a free faunal exchange between these continental fragments; (2) late Cretaceous suturing of the Dutch Antilles and Venezuelan offshore islands to the mainland; (3) oceanic gap remaining between south Central America and northwestern South America; (4) Antillean arc stretched and broken on the eastern end by movement of Caribbean plate (heavy arrow). Broken vector in lower right-hand corner shows movements of South America relative to North America between 65, 38, and 30 Ma.

locity of the Farallon-Cocos-Nazca plate requires that, in Middle America, either there has been extensive Farallon plate consumption on the western margin (similar to the observed late Cretaceous to Recent active margin), or that this plate has been inserted into the Caribbean. I prefer a less mobilist view (Donnelly 1985), which is summarized in Figures 2-2, 2-3, and 2-4. The more mobilist views are presented in Pindell and Dewey (1982) and Wadge and Burke (1983). The controversy between the more and less mobilist views may have only a minor biogeographic significance.

The late Cenozoic eastward movement of the Caribbean plate is itself not controversial. Both the northern and southern margins of the plate are zones of active strike-slip seismicity and neotectonics. The existence of a young spreading center in the Cayman trough (Holcombe et al. 1973) demonstrates

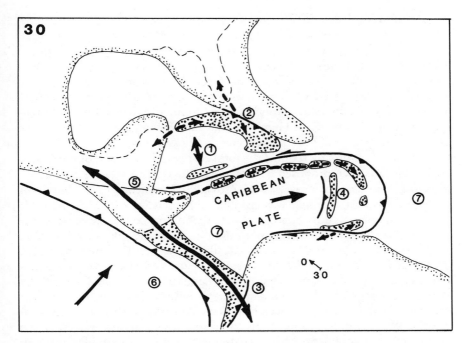

Figure 2-4. Sketch showing possible configuration of continental fragments and island arcs in middle Tertiary, about 30 Ma. Symbols as in previous figures. Numbered points are (1) Yucatán Basin, which opened in latest Cretaceous and early Tertiary, pushing Cuba northward; (2) collision between western Greater Antilles and Bahama platform in early Tertiary; (3) collision between southern Central America and northwestern South America in early Tertiary; (4) Aves ridge, which is apparently an earlier Lesser Antillean island-arc (showing a questionable subduction zone on the west side); (5) the Motagua zone, which shows limited sinistral offset in response to the early opening of the Cayman trough (arrow on north Caribbean fault zone); (6) and (7) Pacific and Atlantic-Caribbean oceanic water masses, which began to develop characteristic differences in silica content of deep and intermediate water masses following the closing of the Central American isthmus. Vector in lower right-hand corner shows movements between 30 and 0 Ma.

a relative velocity of about 2 cm/yr between the Caribbean and North American plates during the past few million years. The extrapolation of this movement back into the middle and early Cenozoic, however, has created problems that are unsettled and may have profound biogeographic significance. Such extrapolation is strongly suggested by the topography of the Cayman trough, which is an elongate parallelogram whose long (east-west) dimension suggests about 1000 km of strike-slip displacement since about 50 Ma, assuming an average velocity of 2 cm/yr. The 1000 km of putative eastward movement of the Caribbean plate presents serious problems in Central America. Careful studies (Burkart 1978) in Guatemala have shown that one of the two potential fault systems across which major post-Cretaceous movements could have taken place (the Polochic fault zone) has only about 130 km of measurable movement. The other possible fault is the Motagua fault zone, and the movement here is apparently extremely limited (Donnelly et al. 1988). If 1000-km movement has taken place, then more than 800-km offset must be accommodated by other means, of which the only two are movement across yet another fault, such as the Jocotán-Chamelecón fault of northwestern Honduras and southern Guatemala, or stretching and crustal thinning of Honduras and the Nicaraguan rise. I prefer the second of these possibilities and develop it below.

A scheme that greatly reduces the total eastward movement of the Caribbean plate has been proposed by Donnelly (1985). This scheme calls attention to the existence of a vast basaltic province (Caribbean "flood" basalt) of middle Cretaceous age that was erupted on preexisting oceanic crust. This igneous complex and its circumferential island arcs constitute most of the modern Caribbean plate. Because of the purely Pacific origin of this plate, fragments of it were emplaced tectonically around the Caribbean during a relatively brief period in the late Cretaceous and early Cenozoic. Emplacement on islands of oceanic origin (Cuba, Jamaica, Hispaniola, Puerto Rico) or on southern Central America are not definitive; these islands and their emplaced bodies could have been transported together with a moving Caribbean plate. However, in several places this igneous complex was emplaced on the edge of continental crust: Guatemala, western Colombia, northern Venezuela, and the islands offshore of Venezuela. Perhaps the most important occurrence from the standpoint of interpretation of Cenozoic movement of the Caribbean plate is the Villa de Cura complex of northern Venezuela, which occupies the Serranía del Interior south of Lake Valencia. This complex is composed of identifiable fragments of the distinctive igneous lithologies of the Caribbean plate and its marginal island arcs, and its major age of emplacement is dated as Paleocene (Donnelly 1985), with the potential of an earlier minor emplacement during the late Cretaceous. During the Paleocene the more mobilist models locate the Caribbean plate far to the west of this site. The identification of Caribbean plate fragments in this complex appears to require that the Caribbean plate was approximately in its present position at least 55 Ma ago.

The Implications of Oceanic Sediment Composition for Terrestrial Biogeography

Oceanic (pelagic) sediment analysis provides an opportunity that has been largely overlooked in the evaluation of tectonic hypotheses. The creation of an emergent "land bridge" would divide a formerly continuous ocean, and the isolation of the oceanic basins may be reflected by differences in accumulating pelagic sediment. Differences in the sediment may reflect either deep- and shallow-water isolation; examples of both are found in the Caribbean. These differences, or the absence of differences, provide strong arguments for or against the existence of emergent bridges.

The study of oceanic sediment during the last two decades of ocean drilling (the Deep Sea Drilling Project) has shown that this sediment reflects distributional patterns in ocean chemistry. Two instances of sediment compositional patterns seen in the Caribbean are responses to circulation of water below about 2000–4000 m depth. Their importance for tectonic or biogeographic analysis is that contrasts in coeval sediments between adjacent basins probably signifies interruption of deep-water flow, which in turn may signify either tectonic closure or the formation of an island arc. While neither of these possibilities necessarily implies total emergence of the barrier, we should note that in the present ocean similar barriers that lack an emergent portion are relatively uncommon (the Walvis ridge is an example).

The first example of deep-water sediment contrast is in biological silica. As is well understood today in marine chemistry, silica migrates in deep water from the Atlantic (the basin of initial reception of most of the world's river water with its dissolved chemical species) to the Pacific in deep-water circulation via the Antarctic Ocean. Brought to the surface in widespread upwelling, it is taken up by radiolaria and diatoms and falls again as siliceous sediment to the ocean floor. The small content of silica in Atlantic deep water results in very limited production of siliceous organisms, and in a notably low-silica sediment (silica does not return from the Pacific but is trapped there; see Broecker 1974 for a concise explanation). When the Central American isthmus was broadly open to the Pacific, there was no barrier to the passage of silica-rich deep water, and early Cenozoic sediments of the central Atlantic, Caribbean, and Pacific have similarly high silica contents. As shown by Donnelly (1985), the closure of Central America between about 38 and 15 Ma is reflected clearly in the upwardly diminishing silica content of a sediment core in the central Venezuelan basin. Thus the Caribbean sediment composition neatly records the timing of the formation of a barrier that could have been surmounted by a chain of islands available for limited terrestrial faunal dispersal between the Americas.

A second example of deep-water connection is seen in late Cenozoic sediments from the Caribbean and adjacent Atlantic, which record a shallowing of the surface at which calcareous microfossil debris dissolves (the "lysocline" or "carbonate compensation depth"; see Broecker 1974). This depth is currently greater than 5000 m in the central Atlantic, but it shallowed throughout the Atlantic during the late Miocene to approximately 4000 m. This event is recorded equally in the central Caribbean and the adjacent Atlantic, which suggests that the Caribbean was then broadly open to the Atlantic, in contrast to the present condition, in which two narrow passages (near the Virgin Islands, and between Jamaica and Haiti) are found

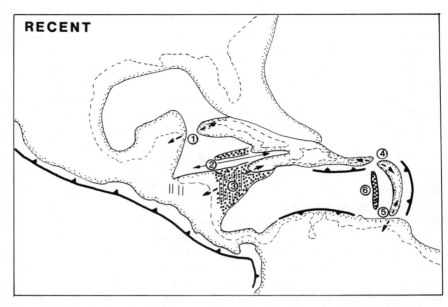

Figure 2-5. Sketch showing the Recent configuration of land masses around the Caribbean. Heavy stippling shows three areas of possible mid-Tertiary or later submergence. Numbered points are (1) Possible faunal connection between the Yucatàn and western Cuba throughout the Cenozoic; (2) Cayman trough opening, with extent of double-headed arrow indicating approximate 1000 km offset since middle Cenozoic; (3) Nicaraguan rise, which is here considered to be stretched and thinned Continental crust of the Chortis block; (4) and (5) the north and south ends of the Lesser Antillean arc, both of which are now separated by deep (about 2000 m) water from adjacent land masses; (6) Aves ridge, which evidently subsided prior to the late Miocene.

at less than 2000 m depth. The implication of a Miocene deep-water connection is that a continuous Greater Antilles–Lesser Antilles–South America land connection is impossible in the late Miocene, and that at least some gaps between islands must have been wider than at present. The most likely potential Miocene gaps are between the Lesser and Greater Antilles (number 4 of Fig. 2-5), and between Grenada and the South American mainland (number 5), but a gap within the Greater Antilles cannot be ruled out. In either case, the potential for dispersal of terrestrial faunas along the Caribbean island chains must have been less than at present.

The closure of the shallow-water connection between the Atlantic and Pacific has long been known, but the data of Keigwin (1982), who analyzed oxygen isotopes in pelagic foraminifera from Caribbean and Pacific sediments, provide the most accurate indication of the time of closure. He found that at 4.1 Ma, the oxygen isotopes on the Caribbean side are consistently heavier than on the Pacific, showing the isolation of shallow-water masses. Shallow-water circulation is limited to the upper hundred meters or so, and this closure indicates essentially complete terrestrial emergence. It was the culmination of a process that began 25 Ma earlier, and it provided the opportunity for North and South American terrestrial faunas to interchange without, literally, getting their feet wet. Yet the connection was probably low (it is now in the order of a few tens of meters in several places) and possibly leaky at times. Thus some shallow marine organisms might have penetrated from one ocean to another, though mollusk and echinoderm faunas imply that this occurred only rarely (Woodring 1965, Durham 1985).

A further implication of Pliocene shallow-water closure is that a large current of westward-directed warm surface water must have been diverted to the north, forming the Gulf Stream, and perhaps causing a dramatic climatic change (an increase in humidity) in the Yucatán-Cuba-Florida area. This, of course, could have immense biogeographic significance.

In summary, deep marine (pelagic) sediment evidence now in hand implies that in the early Cenozoic the Caribbean was broadly open to the Pacific. The connection between the Caribbean and Atlantic was broader and deeper at a more recent time (about 10 Ma), suggesting that terrestrial dispersal among the Lesser Antilles, the Greater Antilles, and South America was more difficult than at present. Finally, the Pliocene shallow-water closure reflects closely the well-known "flood" of interchanging North and South American mammals, and implies that other terrestrial organisms dispersed similarly.

Plate Tectonic Schemes for Caribbean Evolution

The History of the Greater Antilles

Because the Greater Antilles have received more attention in recent biogeographic literature than other parts of the Caribbean, a summary of pertinent geologic observations and their interpretations will be useful.

There is no direct evidence (exposure of bedrock) nor indirect evidence (radiogenic isotope enrichment in igenous rocks) for the existence of old continental crust beneath the Greater Antilles. The oldest rocks in the eastern islands are early Cretaceous oceanic volcanics erupted at abyssal depths. At the western end, the oldest rocks are Jurassic clastic sediments deposited at unknown water depths. It may be assumed, thus, that no island is a fragment of continental material. Cuba, however, remains somewhat of a geologic enigma because of its structural complexity. It appears to consist presently of fragments of several terranes, including part of the Florida-Bahama continental margin block, extensive island-arc sediments and volcanic rocks, thick accumulations of the Cretaceous "flood" basalt, and possible fragments of older continental material derived from the south (e.g., Chortis block). These fragments appear to have been tectonically consolidated during the late Cretaceous with renewed compressive deformation in the Eocene (Pardo

1975). Cuba provides a possible narrow exception to a strictly noncontinental origin of the Greater Antilles.

The first emergent lands in the Greater Antilles appeared in Puerto Rico and the Virgin Islands at about 105 Ma or slightly earlier. Similar rocks of this age are more poorly known in Hispaniola and Jamaica, but we might suppose that similar islands appeared there also. The early island arc appears to have fringed the flood basalt province and is probably causally related to it. La Désirade, which is just east of Guadeloupe, may be a fragment of the early Greater Antilles arc. The islands of Tobago and Bonaire are underlain by very similar rock associations, and a complete arc joining northern Central America and northern South America is easily possible. Such an arc does not imply, however, a pathway by which terrestrial organisms might traverse readily from one end to another, but only a series of persistent volcanic centers, more or less equally spaced, which grow to emergence and provide way stations for filter dispersal of the more vagile elements of the fauna.

Beginning at about 80 Ma the character of Greater Antillean deformation changed markedly. A dramatic increase in the sedimentary component of the rock assemblage implies increasing compressive deformation in an island-arc assemblage. We must assume that at this time the total area of emergent islands increased greatly. It is possible, but not proved, that persistent land connections existed between several of the islands, and also with Central America on the west and South America on the south.

In the early Cenozoic, compressive deformation diminished greatly. In several islands tensional features—deep rifts surrounded by emergent highs—cut across earlier structures. Although the pattern of emergence undoubtedly changed at this time, the total area of emergent land probably did not.

At about the end of the Cretaceous, the formation of the Yucatán Basin swept the fragments of proto-Cuba northward to their present position against the Florida-Bahama block. During their northward travel, fragments of Cuba as well as the Chortis block and the earlier Motagua suture zone were left along the present eastward margin of Belize and Yucatán. Thus, during a period of several million years, western Cuba was probably fairly close to Central America, though there is no evidence for a direct land connection at any time.

At an unknown and disputed time in the Cenozoic—possibly beginning as early as 40 Ma or as late as about 10 Ma—the Greater Antilles, with the possible exception of Cuba, underwent extensive left-lateral strike-slip fault movement. This had the effect of juxtaposing for varying lengths of time east-west oriented slivers of different islands. A possible secondary effect was a scissor motion on some of the faults, with some blocks rotated about a horizontal north-south axis so that some parts were submerged and other parts emerged.

It is noteworthy that there are no existing stratigraphic elements of the

Cretaceous and early Cenozoic of any Greater Antillean island that are peculiar to that island, and that even seemingly unusual rock types are commonly found on more than one island. Instead, one finds striking similarities between the late Cretaceous limestones of southern Haiti, for example, and Jamaica; between the early Cenozoic conglomerates of eastern Cuba and Jamaica; between middle and late Cretaceous alkalic volcanic rocks of Hispaniola and Puerto Rico. Perhaps the best way to describe the Greater Antilles is not as four major islands but as about a dozen blocks that have at different times been juxtaposed and wrenched apart, generally with a sinistral sense of offset of the elongate parts. The total result has been the lengthening of the island chain with relative brief submergence or emergence of the depressed zones between these blocks. The shallow water between southern Haiti and Jamaica, for example, has no more permanence as a water barrier than the Enriquillo Valley between southern and central Hispaniola has as a terrestrial connection.

The principles of isostasy imply, and the stratigraphic record supports, that the largest and highest elevation of the blocks probably has had the lowest probability, or briefest time, of complete submergence. Jamaica, which is relatively low standing, for example, seems to have been completely submerged during part of the Oligocene (White Limestone Formation; Arden 1975). Stratigraphic evidence for similar submergence for other major portions of the Greater Antilles is unclear, but the general tectonic relaxation during the middle Cenozoic would be the time interval favored for such submergence if it occurred. On most islands there is a sufficient record from younger sediments to suggest that a coastal plain surrounded emergent islands during most of the middle and late Cenozoic.

The Nicaraguan rise between Honduras and Jamaica is a major feature of the western Caribbean that is still poorly understood. Its water depth, seismic refraction velocities (Edgar et al. 1971), and gravity all suggest that it is basically continental and an eastward prolongation of the Chortis block. The poorly defined topography of the shallow axis of the rise suggests north-south grabens. The junction with Jamaica is not understood, but it might represent a compressive suturing feature. This rise may have been an emergent feature stretched into submergence during middle and late Cenozoic time. Such a stretching, shown in Figure 2-5, implies differing modes of deformation on the northern and southern sides of the north Caribbean fault zone, with the total offset largely taken up by stretching of the southern side, that is, by an internal deformation of the Caribbean plate. Also consistent with this mode of deformation is the widespread distribution in the northwest Caribbean (eastern Costa Rica and Isla Providencia to Honduras and central Hispaniola) of Pliocene and Pleistocene alkalic and subalkalic volcanic centers that are most commonly located along approximately north-south normal faults (Wadge and Wooden 1982, Walker 1981).

There are two potential terrestrial connections between Central America and the Greater Antilles during the Cenozoic. Shallow-water extensions of Cuba and the Yucatán are currently about 150 km apart; the geologic framework of the Belize-Yucatán border suggests that this distance has never been larger and may have been smaller during the early Cenozoic opening of the Yucatán basin. A second connection might have been between Honduras and Jamaica via the now submerged Nicaraguan rise. Although shallow-water areas are currently about 350 km apart, a prestretching configuration might have provided a continuous terrestrial connection. This connection could serve as Rosen's (1975) vicariant pathway rather than the more speculative proto-Greater Antilles extending mysteriously from the southern Chortis block. The timing of these two potential connections would be different: the Cuba-Yucatán connection might never have been a persistent terrestrial connection, but would have remained at approximately the same distance throughout the Cenozoic. The Honduras-Jamaica connection would have been severed dramatically during the middle Cenozoic because of the subsidence of the Nicaraguan rise.

The eastern and southern extension of the Greater Antilles is problematical. The similarity between the volcanic series and shallow plutonic rocks of the British Virgin Islands and St. Martins–St. Barts strongly suggests an early Cenozoic connection between these areas. The island of La Désirade, just east of Guadeloupe, is underlain by early Cretaceous volcanic rocks nearly identical to those of St. Thomas and St. John. Finally, the islands of Tobago and Bonaire are underlain by Cretaceous volcanic rocks similar to those of the Virgin Islands and Puerto Rico. Although these fragments cannot be connected in a convincingly clear island chain, it appears that a once-continuous Cretaceous island arc has, during the Cenozoic, been stretched into a series of disconnected fragments. The longest gap in which there is no present evidence of this arc in between Guadeloupe and Tobago, about 550 km.

A Cretaceous proto-Greater Antilles provided the best possibility for an early dispersal pathway between the Americas. Cenozoic tectonic movements resulting from eastern movement of the Caribbean plate have probably torn this arc in its eastern portion while juxtaposing diverse slivers in its northern portion.

The Geologic History of the Lesser Antilles

The topography of the Lesser Antilles is so dramatically dominated by a curvilinear chain of active volcanoes that it is easy to forget that these features are relatively quite young and possibly of limited paleobiogeographic significance. Built at least in part on fragments of the Greater Antilles, from the Miocene to the Recent, volcanic centers developed in essentially their present locations, although in some islands, the older centers are identifiable some kilometers from the Recent centers. There appear to have been no strike-slip fault movements parallel to the arc, unlike the Greater Antilles, and there is no evidence that slivers were displaced along the length of the arc. Although some vertical oscillations are possible, there is evidence against large vertical movements of any of the islands; rather, they simply have grown slowly upward through accretion of volcanic materials, always maintaining a near-isostatic balance. Water depths are too great for Pleistocene emergent connections between these young islands, especially in the center of the arc. Furthermore, there is no evidence that volcanic centers existed on presently water-covered, shallow portions of the arc. Thus the sum of the observations is that there could have been no continuous land connections between the volcanic islands at any stage of their evolution from the Miocene onward.

Connections of the Lesser Antilles with Puerto Rico and the Virgin Islands platform have been suggested by some researchers. The Anegada trough separating the Greater and Lesser Antilles is very deep, with a sill depth at about 2000 m and extensive basins with water depths greater than 4000 m. Although it grossly resembles a tensional feature, its orientation with respect to the northwestward movement of the South American plate suggests compression in this orientation. Furthermore, it is a nearly direct eastward extension of the clearly compressive subduction of the Muertos trough, south of Puerto Rico and Hispaniola. The conclusion that the trough represents a relatively young graben developed on a continuous platform is overly simplistic; a more realistic view is that it is dominantly a strike-slip feature with associated tensional basins. The age of its formation is not known.

Past connections between the southern Lesser Antilles (Grenada) and the older island-arc segments adjacent to South America (Tobago, Venezuelan offshore islands) are not known. The history of this area has recently been discussed by Speed (1986), who concludes that the southern Lesser Antilles developed in the Cenozoic as a southeast-facing island arc located north of South America. Later northward movements of South America resulted in a collision between the island-arc and the mainland. An inference drawn from Speed's interpretation is that throughout the Cenozoic the Lesser Antilles was farther from South America than at present, and that an emergent link between the two was not likely.

The enigmatic Aves ridge is a linear submerged feature between the Greater Antilles and South America about 200–300 km west of the Lesser Antilles. In the early Cenozoic it was emergent (Bouysse et al. 1985), but it has now subsided to a minimum depth of about 1 km. This feature may be a part of the early and middle Cenozoic island arc mentioned above, or it might be yet another arc. The dredging of shallow-water sediments on the ridge suggests that it provided a possibly redundant alternative for terrestrial faunal movements during the early Cenozoic, but it is unlikely that it was connected to South America.

In summary, in contrast with the Greater Antilles, the Lesser Antilles appears to be a relatively young and uncomplicated series of emergent volcanic centers, developed on and largely obscuring dispersed fragments of an older island arc. Biogeographically, these islands offer limited opportunity for dispersal from South America, and from their geologic history one would suspect an older faunal association with the Greater Antilles and only a very limited association with South America.

Implications for Caribbean Biogeography

A reevaluation of Caribbean tectonic history in light of the plate-tectonic revolution of the 1960s has suggested many new approaches to the problem of dispersal of terrestrial faunas. At the same time, we have strengthened older beliefs as to the relative permanence of emergent or submergent areas in light of the principle of isostasy. Many problems—the major one being the lack of certain knowledge of the early movement of the Farallon plate—are still unsolved.

The earliest stage of opening between the Americas in the Jurassic and early Cretaceous isolated Gondwanaland from the northern continents. The eruption of the "flood" basalts of the early and middle Cretaceous were apparently accompanied by the development of a circumferential island arc on at least the eastern side (Fig. 2-2). This arc can serve as a filter-bridge pathway for the limited terrestrial dispersal required in late Cretaceous by mammalian and reptilian fossil evidence. It also serves as Rosen's (1975) vicariant link between the Americas. An important difference between my view and that of Rosen and his geologic predecessors is that his view implies that this link may have later become an elongate, isolated island in the Caribbean, while my view implies that it was always more or less linked at the northwestern end.

The beginning of inter-American plate compressive movements in the late Cretaceous marked the strong development of two island-arc systems, the Greater Antilles and southern Central America, which, of course, later became totally emergent and linked as an intercontinental isthmus. Vicariant opportunities within the arcs increased with emergence (Fig. 2-3). However, the intercontinental link had been severed with eastward movement of the Caribbean plate, which stretched and severed the eastern end of the proto-Antillean arc. Cuba formed approximately at the end of the Cretaceous; its weak link with the Yucatán peninsula continued, and a new link was forged with the Florida-Bahama block. The suturing of the Maya and Chortis blocks ended a period of isolation for the latter; terrestrial pathways from North America to the Greater Antilles may also have included a link via

Honduras and Jamaica. The suturing of the remnants of the proto-Antilles with South America in the late Cretaceous probably has limited biogeographic significance because this link did not extend northward.

In the middle Cenozoic, compressive interplate relaxation greatly diminished the construction of the Greater Antilles island arc and major tensional features led to formation of important grabens and subsidence of many of its fragments (Fig. 2-4). This was a time both of isolation of many of the smaller islands and of diminished dispersal among the islands. At the same time, the eastward movement of the Caribbean plate closed the isthmian connection of Central America. Fossil evidence suggests that the barrier was sparsely surmounted with islands, but several vertebrate groups evidently dispersed between the Americas.

The opening of the Cayman trough in the early or middle Cenozoic is the most spectacular manifestation of the increasing eastward movement of the Caribbean plate. Strike-slip movements were distributed among an anastomosing series of shears within larger blocks of the Greater Antilles, and the effect was to shift these blocks with respect to one another. The present grouping of blocks into large islands and the connecting emergent or submergent zones between the blocks are both geologically temporary.

The strike-slip movement of the Cayman trough was apparently largely expressed as internal deformation of the northwestern Caribbean. The most significant manifestation of this stretching was the apparent subsidence of the Nicaraguan rise, which left Jamaica isolated from Honduras during the middle Cenozoic.

In the eastern Caribbean, fragmentation and isolation of the proto-Antillean arc continued well into the Cenozoic. A younger south-facing arc within the Caribbean developed in the early Cenozoic, in part possibly on fragments of the older arc. In the middle and late Cenozoic the eastern Caribbean was broadly open to the Atlantic.

In the late Cenozoic, eastward movement of the Caribbean plate was accompanied by a young volcanic arc, which today forms the bulk of the Lesser Antilles. This young arc approached South America during this time. Thus the terrestrial connection between the Greater Antilles and South America is probably closer than at any time since the fragmentation of the proto-Antilles at the end of the Cretaceous. Terrestrial faunal distributions of the eastern Caribbean should reflect the recency of this connection. There is no evidence supporting notions that important emergent segments within the Lesser Antilles have been uplifted sufficiently to create young terrestrial connections among the islands. Though the distribution of some montane species might seem to require continuous high-elevation ridges in the Lesser Antilles where water now separates volcanic centers, geologic results increasingly reinforce the simpler view that such ridges cannot have existed between present young volcanic islands, and that any valid example of such faunal distributions must seek its explanation in a middle Cenozoic or older connection within these islands.

The Problem of Paleoclimatology and Caribbean Biogeography

The problem of past climates has received remarkably little attention from Caribbean biogeographers. We can surmise from the widespread development of lateritic soils in the Greater Antilles that, during the middle Cenozoic, climates must have been less moderate than at present and must have been marked by strongly alternating wet and dry periods. The effect of the creation of the Gulf Stream coincident with the closing of the Central American isthmus during the Pliocene must have had profound climatic implications for the western Caribbean and the southeastern United States, but remarkably little paleontologic documentation of such changes exists.

The glacial periods have received a great deal of attention recently, mainly from a group of marine geologists and meteorologists known as CLIMAP (1976). In the Caribbean there is strong evidence for aridity during periods of glacial maxima, including evidence from Cuban soils, from eolian deposits of Puerto Rico and the Dutch Antilles, and from evidences for glaciation on Hispaniola. These have been summarized recently by Schubert (in press). Any account of the terrestrial biogeography must include a strong consideration of these implications; however, our present evidence is woefully inadequate for the construction of a suitable climatic history for any part of the Caribbean region.

Summary

The continuous America-Europe-Africa land mass was fragmented beginning in the Triassic. North and South America were separated starting in the early Jurassic with oceanic crust filling the intervening space. In the early Cretaceous, Africa separated from South America, and at that time the generation of oceanic crust between the Americas changed into a "flood" basalt magmatism (Fig. 2-1). A circumferential island arc connected the two Americas tenuously and provided a limited terrestrial connection between the Americas. This is reminiscent of the proto-Greater Antilles of several authors that was presumed to have originated at the present site of Central America and later migrated into its present position. The beginnings of tectonic compression between the Americas at about 80 Ma intensified the development of the island-arc system and built the island chains into prominent emergent features.

In the latest Cretaceous and early Cenozoic this island arc was stretched and finally broken into fragments as the flood basalt (now identifiable as the Caribbean plate) moved eastward. Tectonic relaxation of inter-American compression resulted in local foundering of Antillean islands, with some smaller blocks (such as the Jamaican block) submerged for significant lengths of time. During this period South America was isolated, and terrestrial faunas on Antillean islands entered into their nadir of local isolation.

The locations of connections between the proto-Antilles and Central and South America have not been identified; such connections can only be presumed to have existed because vertebrate fossil evidence requires them. In the late Cretaceous the opening of the Yucatán Basin swept diverse terranes northward, forming Cuba. The western end of Cuba may be presumed to have been sufficiently proximate to the Maya block (Yucatán) to serve as a continuing filter bridge. Following the late Cretaceous suturing of the Chortis and Maya blocks, a terrestrial connection between Honduras and Jamaica may also have existed via the Nicaraguan rise.

In the middle Cenozoic (Oligocene to early Miocene) the continued eastward movement of the Caribbean plate closed the deep oceanic connection across the Central American isthmus. Vertebrate fossil evidence suggests that the island arc of Costa Rica and Panama served as a limited filter bridge for terrestrial fauna migration.

Continued eastward movement of the Caribbean plate was manifested on the north with the opening of the Cayman trough, which required about 1000 km of sinistral movement commencing at an unspecified time in the early Cenozoic and continuing until the present. If the response to this offset was stretching of the Caribbean plate, then a submergence of the Nicaraguan rise and a late Cenozoic removal of this terrestrial connection is highly plausible. Eastward movement of the Caribbean plate also had the effect of stretching the Greater Antilles and moving blocks within it into temporary juxtaposition with or separation from adjacent blocks. The valleys marking the courses of the major strike-slip faults were nonpersistent submergent or emergent features.

The Lesser Antilles during the middle Cenozoic is a series of separated fragments more distant from South America than from the Greater Antilles. The volcanic arc formed during the late Cenozoic now provides a filter bridge for dispersal from South America, but geologic evidence minimizes an earlier Cenozoic connection.

For most of late Mesozoic and Cenozoic time, faunal movements would have required short (tens of kilometers) overwater dispersal. During relatively brief intervals—late Cretaceous through a proto-Antillean arc and middle Cenozoic through Central America—there might have been relatively narrow water gaps. Faunal movements within the Greater Antilles, including one or two connections with northern Central America, could have been terrestrial for brief intervals and never required long overwater dispersals. Faunal movements within the Lesser Antilles, on the other hand, appear to have always required overwater dispersals, at least from the middle Cenozoic onward.

Thus, as the older notion of emergent oceanic areas has had to be abandoned, a newer view that terrestrial land bridges were important and possibly persistent features of faunal dispersal in the Caribbean appears to be equally doomed in the face of increasing geologic, especially marine-geologic, information. Our present evidence seems to place a far greater emphasis on overwater dispersal at the expense of vicariance, but geologic evidence is beginning to find limited possibilities for vicariant faunal exchange.

Acknowledgments

The Caribbean biogeography symposium at the Annual Meeting of the Entomological Society of America in Hollywood, Florida (December 1985) provided its participants with a powerful stimulus to explore interdisciplinary problems pertinent to this complex topic, and I am grateful to Jim Liebherr for inviting me to attend. A more recent meeting of the 11th Geological Conference in Barbados (July 1986) provided additional valuable insights, and I am especially grateful to Bob Speed and Carlos Schubert for relevant ideas.

Note. Subsequent to the preparation of this chapter, I undertook additional analyses of Atlantic pelagic sediment and have extended the conclusions I reported. Pelagic sediments from the Atlantic show sharply diminished biological silica at 45 Ma and essentially no biological silica after 35 Ma. In contrast, Caribbean pelagic sediments record biological silica until 15 Ma. I conclude that an effective barrier to deep water circulation came into existence at about 45 Ma, probably at the site of the Aves Ridge. This barrier would have been an island arc, which would have served as a filter bridge for dispersal from South America to the Greater Antilles. As discussed in this chapter, the barrier must have been breached by the late Miocene, when the Atlantic lysocline excursion was recorded in the Caribbean. A fuller discussion of this subject will appear in "Biogeography of the West Indies," ed. C. Woods, E.J. Brill, Netherlands.

References

Anderson, T.H., and V.A. Schmidt. 1983. The evolution of Middle America and the Gulf of Mexico—Caribbean Sea region during Mesozoic time. Geol. Soc. Amer. Bull. 94:941–966.

Arden, D.D., Jr. 1975. Geology of Jamaica and the Nicaragua Rise. *In* A.E.M. Nairn and F.G. Stehli, eds., The Ocean Basins and Margins, vol. 3: The Gulf of Mexico and the Caribbean, pp. 617–661. Plenum Press, New York.

Barr, K.W. 1974. The Caribbean and plate tectonics—some aspects of the problem. Verhandl. Naturf. Ges. Basel 84:45–67.

Beets, D.J., W.V. Maresch, G. Klaver, A. Mottana, R. Bocchio, F.F. Beunk, and H.P. Monen. 1984. Magmatic rock series and high-pressure metamorphism as constraints on the tectonic history of the southern Caribbean. In W.E. Bonini, R.B. Hargraves, and R. Shagam, eds., The Caribbean–South American Plate Boundary and Regional Tectonics, pp. 95–130. Geol. Soc. Amer. Mem. 162.

Bouysse, P., P. Andreieff, M. Richard, J.C. Baubron, A. Mascle, R.C. Maury, and D. Westercamp. 1985. Aves Swell and northern Lesser Antilles Ridge: Rock dredging results from Arcante 3 cruise. In A. Mascle, ed., Geodynamique des Caraïbes. Symposium, Paris, pp. 65–76. Edit. TECHNIP, Paris.

Broecker, W.S. 1974. Chemical Oceanography. Harcourt Brace Jovanovich, New York. 214 pp.

Burkart, B.B. 1978. Offset across the Polochic fault of Guatemala and Chiapas, Mexico. Geology 6:328–332.

Burke, K., C. Cooper, J.F. Dewey, P. Mann, and J.L. Pindell. 1984. Caribbean tectonics and relative plate motions. In W.E. Bonini, R.B. Hargraves, and R. Shagam, eds., The Caribbean–South American Plate Boundary and Regional Tectonics, pp. 31–63. Geol. Soc. Amer. Mem. 162.

Buskirk, R.E. 1985. Zoogeographic patterns and tectonic history of Jamaica and the northern Caribbean. J. Biogeogr. 12:445–461.

CLIMAP. 1976. The surface of ice-age earth. Science 191:1131–1137.

Coney, P.J. 1982. Plate tectonic constraints on the biogeography of Middle America and the Caribbean region. In N.R. Morin, ed., Biological Studies in Central America, Twenty-eighth Annual Systematics Symposium. Ann. Missouri Bot. Gard. 69:432–443.

Dickinson, W.R., and P.J. Coney. 1980. Plate tectonic constraints on the origin of the Gulf of Mexico. In R.H. Pilger, Jr., ed., The Origin of the Gulf of Mexico and the Early Opening of the Central North Atlantic Ocean, pp. 27–36. Louisiana State University, Dept. of Geology.

Donnelly, T.W. 1985. Mesozoic and Cenozoic plate evolution of the Caribbean region. In F.G. Stehli and D. Webb. eds., The Great American Biotic Interchange, pp. 89–121. Plenum Press, New York.

Donnelly, T.W., G.S. Horne, R.C. Finch, and E. López-Ramos. 1988. Northern Central America. In J. Case and G. Dengo, eds., The Geology of North America, Vol. H, The Caribbean Region. Geol. Soc. Amer.

Duncan, R.A., and R.B. Hargraves. 1984. Plate tectonic evolution of the Caribbean region. In W.E. Bonini, R.B. Hargraves, and R. Shagam, eds., The Caribbean–South American Plate Boundary and Regional Tectonics, pp. 81–93. Geol. Soc. Amer. Mem. 162.

Durham, J.W. 1985. Movement of the Caribbean Plate and its importance for biogeography in the Caribbean. Geology 13:123–125.

Edgar, N.T., J.I. Ewing, and J. Hennion. 1971. Seismic refraction and reflection in the Caribbean. Amer. Assoc. Petrol. Geol. Bull. 55:833–870.

Freeland, G.L., and R.S. Dietz. 1971. Plate tectonic evolution of the Caribbean region. Nature 232:20–23.

Holcombe, T.L., P.R. Vogt, J.E. Matthews, and R.R. Murchison. 1973. Evidence for sea-floor spreading in the Cayman Trough. Earth and Planetary Sci. Lett. 20:357–371.

Keigwin, L.D., Jr. 1982. Isotopic paleoceanography of the Caribbean and east Pacific: Role of Panama uplift in late Neogene time. Science 217:350–353.

Ladd, J.W. 1976. Relative motion of South America with respect to North America and Caribbean tectonics. Geol. Soc. Amer. Bull. 87:969–976.

MacDonald, W.D. 1976. Cretaceous-Tertiary evolution of the Caribbean Transactions, 7th Conf. Geol. Caraïbes (Guadeloupe), pp. 69–78.

MacFadden, B.J. 1981. Comments on Pregill's appraisal of historical biogeography of Caribbean vertebrates: Vicariance, dispersal, or both? Syst. Zool. 30:370–372.

MacGillavry, H.J. 1970. Geological history of the Caribbean. Kon. Ned. Akad. Wet. 73 (ser. B):64–96.

Malfait, B.T. and M.G. Dinkleman. 1972. Circum-Caribbean tectonic and igneous activity and the evolution of the Caribbean plate. Geol. Soc. Amer. Bull. 83:251–272. Discussion by Meyerhoff and Meyerhoff, and reply, 84:1105–1108.

Mattson, P.H. 1984. Caribbean structural breaks and plate movements. In W.E. Bonini, R.B. Hargraves, and R. Shagam, eds., The Caribbean–South American Plate Boundary and Regional Tectonics, pp. 131–152. Geol. Soc. Amer. Mem. 162.

Meyerhoff, A.A. 1967. Future hydrocarbon provinces of the Gulf of Mexico–Caribbean region. Trans. Gulf Coast Assoc. Geol. Soc. 17:217–260.

Meyerhoff, A.A., and H.A. Meyerhoff. 1972. Continental drift: IV, The Caribbean "Plate". J. Geol. 80:34–60.

Nairn, A.E.M., and F.G. Stehli, eds. 1975. The Ocean Basins and Margins, vol. 3: The Gulf of Mexico and the Caribbean. Plenum Press, New York. 706 pp.

Pardo, G. 1975. Geology of Cuba. In A.E.M. Nairn, and F.G. Stehli, eds., The Ocean Basins and Margins, vol. 3: The Gulf of Mexico and the Caribbean, pp. 553–615. Plenum Press, New York.

Perfit, M., and B.C. Heezen. 1978. The geology and evolution of the Cayman Trough. Geol. Soc. Amer. Bull. 89:1155–1174.

Pindell, J.L. 1985. Alleghenian reconstruction and subsequent evolution of the Gulf of Mexico, Bahamas, and proto-Caribbean. Tectonics 4:1–39.

Pindell, J., and J.P. Dewey. 1982. Permo-Triassic reconstruction of western Pangea and the evolution of the Gulf of Mexico/Caribbean region. Tectonics 1:179–211.

Pregill, G.K. 1981. An appraisal of the vicariance hypothesis of Caribbean biogeography and its application to West Indian terrestrial vertebrates. Syst. Zool. 30:147–155.

Rosen, D.E. 1975. A vicariance theory of Caribbean biogeography. Syst. Zool. 24:431–463.

Schmidt-Effing, R. 1980. The Huayacocotla aulacogen in Mexico (lower Jurassic) and the origin of the Gulf of Mexico. In R.H. Pilger, Jr., ed., The Origin of the Gulf of Mexico and the Early Opening of the Central North Atlantic Ocean, pp. 79–86. Louisiana State University, Dept. of Geology.

Schubert, C. In press. Late Pleistocene glacial aridity of northern South America and the Caribbean. *In* J. Rabassa, ed., Quaternary of South America and the Antarctic Peninsula. Balkema, Rotterdam.

Speed, R. 1986. Cenozoic tectonic evolution of the southeastern Caribbean (abstract), pp. 106–107. 11th Caribbean Geological Conference, Barbados, July 1986.

Stehli, F. G., and S.D. Webb, eds. 1985. The Great American Biotic Interchange. Plenum Press, New York. 530 pp.

Sykes, L.R., W.R. McCann, and A.L. Kafka. 1982. Motion of Caribbean plate during the last 7 million years and implications for earlier Cenozoic evolution. J. Geophys. Res. 87:10656–10676.

Tarling, D.H. 1980. The geologic evolution of South America with special reference to the last 200 million years. *In* R.L. Ciochon and A.B. Chiarelli, eds., Evolutionary biology of the New World Monkeys and Continental Drift, pp. 1–41. Plenum Press, New York.

Wadge, G., and K. Burke. 1983. Neogene Caribbean plate rotation and associated Central American tectonic revolution. Tectonics 2:633–643.

Wadge, G., and J.L. Wooden. 1982. Cenozoic alkaline volcanism in the northwestern Caribbean: Tectonic setting and Sr isotopic characteristics. Earth and Planetary Sci. Lett. 57:35–46.

Walker, J.A. 1981. Petrogenesis of lava from cinder come fields behind the volcanic front of Central America. J. Geol. 89:721–729.

Weyl, R. 1966. Geologie der Antillen. Gebrüder Borntraeger, Berlin. 410 pp.

———. 1973. Die paläogeographische Entwicklung Mittelamerikas. Zent. Geol. Paläont. (1) 5/6 (1973):432–466.

———. 1980. Geology of Central America. Gebrüder Borntraeger, Berlin. 371 pp.

Woodring, W.P. 1954. Caribbean land and sea through the ages. Geol. Soc. Amer. Bull. 65:719–732.

———. 1965. Endemism in middle Miocene Caribbean molluscan faunas. Science 148:961–963.

3 · Zoogeography of West Indian Lygaeidae (Hemiptera)

James A. Slater

Biogeographic analysis of the family Lygaeidae can be valuable for a number of reasons. Lygaeids are relatively numerous as individuals as well as in number of species. They occupy a novel trophic position in that most species feed on mature seeds yet have piercing-sucking mouth parts. A considerable proportion of the fauna lives in ground litter where, under stable conditions, they frequently become flightless with resultant limited dispersion ability. They are an old group. Popov (1981) noted that the Lower Cretaceous fauna of Siberia, Mongolia, and North China is dominated by the great number of Lygaeidae.

Most of those who work with insects, except the fortunate ones who work on butterflies, some beetles, or mosquitoes, face the same problems of negative evidence. For a great many islands of the Antilles (almost all of the very small ones) our knowledge is so inadequate that any attempt to quantify the relationships and interpret such scanty information would be quite unprofitable and grossly misleading. We have a reasonable understanding of the diversity of the faunas of the Greater Antilles and of most of the major islands of the Lesser Antilles from Guadeloupe south through Grenada, and also of Tobago and Trinidad on the South American continental shelf. Unfortunately, inadequate knowledge of distributions in Mexico, Central America, and northern South America severely limits our ability to understand relationships of West Indian Islands to mainland areas for several elements of the West Indian fauna. Rosen (1985) stated that "all problems in comparative biology begin with a data gathering phase that incorporates unknown amounts of noise and signal." Lygaeid studies are to a considerable extent in this phase.

Materials and Methods

The data given in the following tables and in the text frequently will not coincide exactly with those in the literature. I have included some undescribed species; in a few cases I have retained questionable but possible literature records, and in other cases I have excluded published records that appear to be based on misidentifications. Authors' names of species are not included in the text but may be found in the checklist at the end of the chapter.

Figures exclude Tobago and Trinidad because both are on the South American continental shelf and thus cannot really be considered part of the Antillean fauna. I shall, however, refer to their faunas for comparative purposes. Although southern Florida may, with some justification, be considered part of the West Indian fauna, its species are not included in the endemicity tables. There is, however, a brief discussion of Floridian–West Indian relationships.

Vicariance or Dispersal

The question of whether major distribution patterns in the Caribbean have resulted from vicariance or dispersal events is at present a matter of lively debate. The lygaeid fauna of the islands is not sufficiently well known distributionally or cladistically to be able to contribute much to this debate within the islands themselves. Of greater importance is the question of vicariance as an explanation of island-mainland relationships. Rosen (1985) reiterated his belief that dispersalist concepts are worthless by such statements as "some a priori notion that all distributions might be explained by guesswork liberally laced with dispersalist intuition is to ensure that future generations of biogeographers will regard such proposals lightly." Rosen seems to believe that a series of congruent disjunct distributions (we will assume the cladistic relationships have been correctly resolved, which, of course, will often not be the case) are the result of fragmentation of a previously nonfragmented range (vicariance). Actually, if wind patterns, ocean currents, similarity of habitats, and relative proximity of areas persist over a reasonable period of time, congruent patterns could be developed by dispersal as well as by vicariance. It is true that given a geologic hypothesis, one can "test" that hypothesis by vicariance but not by dispersal. Unfortunately, because of many such hypotheses for the Caribbean, vicariance "tests" will be, at the most, possibilities. Rosen's belief that "theories of dispersal to explain biotic complexity are no more informative than theories of relationship based on symplesiomorphy" seems unreasonable to me. It is one thing to recognize the basic importance of three-taxon cladistic statements but quite another to use them as evidence of a vicariant versus dispersal explanation of the pattern one has. Rosen does not really address the important question of *what is not there* and thus forces himself into an almost ritualistic method of interpreting his data. If Rosen is correct and there are 19 resolved four-area cladograms for the Caribbean, then he is also correct in saying that it "will indeed require a stupendous multidisciplinary effort to resolve decisive patterns for the region."

As discussed below, there are a few taxa of Lygaeidae in the West Indies whose distributions may represent ancient mainland-island vicariance. There is such remarkable variance in the timing of Caribbean plate motion at the moment that discussion relative to lygaeid distribution does not seem useful.

Matile (1982) believed that flies of the family Keroplatidae on the Lesser Antilles show a vicariance pattern resulting from a period when these islands lay between North and South America, which would date the fauna from the late Cretaceous or early Eocene. However, if Coney (1982) was correct, there is no evidence of the existence of the Lesser Antilles at this time. Matile's belief that "most of the tracks obtained join the Lesser Antilles to the Greater Antilles, Central America or the Northern Andes" does not indicate to me evidence of a vicariant origin.

It is evident that the Greater Antilles lay closer to Central America during the Tertiary than they do now. The evidence is strengthened by the recent discovery in amber of an army ant of the genus *Neivamyrmex* from Hispaniola, dating either from the late Oligocene or early Miocene age (Wilson 1985). According to Wilson, army ants appear to be among the least vagile of insects (possibly an overstatement), and they do cross small water gaps, occurring in fact on St. Vincent in the Lesser Antilles.

The Lygaeid Evidence

That the West Indian lygaeid fauna is in part the result of dispersal from the mainland is not merely an intuitive belief, as Rosen (1985) would have it. There is some evidence to support this fact.

The following three examples from southern Florida may be instructive in showing that dispersal occurred in the Antilles. First, there is *Heraeus triguttatus*, a brightly colored orange and black lygaeid geophile that is part of a mimetic complex on Cuba. This complex has a *Solenopsis* ant as a model, and the mimics include two additional lygaeids (a species of *Pamphantus* and one of *Myodocha*), a spider, and a beetle. Of these only *H. triguttatus* is known to occur outside of Cuba.

H. triguttatus is common in extreme southern Florida. All other members of the genus *Heraeus* are various shades of dull brown, gray, or black. To believe that *H. triguttatus* did not disperse across water from Cuba after evolving its present mimetic color pattern would require explaining why none of the other mimics, nor the model, is now present either in Florida or anywhere else on the North or South American mainland.

The second example is perhaps less compelling, but nevertheless it suggests that probability must play a part in the interpretation of why things are where they are. *Craspeduchus pulchellus* is a common and widespread species that occurs almost throughout the West Indies. It is a brightly colored insect that lives on *Corchorus siliquosus*, a plant that grows in early succession stages particularly along roadsides. Even though collectors have been on the Florida Keys for many years, it was not taken until 1969 on Key Largo, by a member of R.M. Baranowski's laboratory. The insect is now well established on the Keys but has not yet been taken on the mainland (see Baranowski and Slater 1975).

The third example, *Ozophora laticephala*, is an abundant insect in the Greater Antilles, where it appears to be an obligatory seed predator on fallen seeds of *Ficus*. It was first taken at Key West in 1972 (Slater and O'Donnell 1979) and has subsequently been found breeding at the Homestead Florida Experiment Station. Based upon its use of temporary habitats, the propensity of species of *Ozophora* to fly to light, and the widespread use of light traps in southern Florida primarily to collect Lygaeidae, it was probably introduced not long before 1972.

Using the above examples as evidence, I have adopted Brown's (1978) methodology of analyzing each taxon for evidence, or lack of evidence, of dispersal and colonizing ability. I suggest combining this method with careful cladistic analysis in order to reach defendable conclusions. Just as Rosen (1985) would have us defend our conclusions on the basis of biological data even in the face of conflicting geologic data, so I would suggest that we should attempt to defend the evidence from the taxa under investigation and not be overly influenced by their degree of congruence or noncongruence with other taxa. Cause and effect are subtle things in biogeography. It may be wise not to assume, as Tolstoy has suggested, that when the great bell in the cathedral tower tolls that it is because both hands of one's watch point to twelve.

Endemicity

Brown (1978) recognized four levels of endemicity in the West Indian butterfly fauna: family, genus, species, and subspecies. He noted that endemicity is absent at the family level. This is also true of the Lygaeidae if we substitute subfamily for family.

Table 3-1. Lygaeidae in the Greater Antilles

Island	Total no. of species	Percentage of West Indian fauna	Total no. of endemic species	Percentage of endemic island fauna
Cuba	84	50	23	27
Hispaniola	57	34	13	23
Jamaica	58	34	7	12
Puerto Rico	43	25	5	12
Caymans	18	11	2	11
Dominica	40	24	1	2.5
Grenada	39	23	1	2.5
Guadeloupe	37	22	0	0
St. Vincent	34	20	1	3
Martinique	25	15		
St. Lucia	13	8		
Bahamas	40	24	0	0
New Providence	24	14		
Abaco	22	13		
Bimini	16	9		
Eleuthera	13	8		
Andros	12	7		
Grand Bahama	11	6.5		
Cat	11	6.5		
Mayaguana	11	6.5		
Crooked	10	6		
Great Inagua	9	5		
Long	8	5		
Berry	6	3.5		
Rum Key	5	3		
San Salvador	4	2		
Darby	1			
Turks and Caicos	8	5		
Virgins	19	11		
Upper Leewards	18	11		
Antigua	15	9		
St. Eustatius	3	2		
St. Maarten	3	2		
Saba	3	2		
Monserrat	2	1		
Anguilla	1	—		
St. Kitts	1	—	1	100
Barbuda	1	—		
Miscellaneous				
Mona Island	15	9	1	7
Barbados	9	5		
I. Pines	8	5		

Generic endemicity is very low; of the 52 lygaeid genera known from the islands, only four (8%) are endemic there. These consist of one flightless monotypic ant mimic, *Abpamphantus*, from Cuba; *Neopamphantus* in the same subfamily from Cuba and Hispaniola; and two Antillocorini—*Antillodema* in the Greater Antilles and *Bathydema*, ranging from the Greater Antilles to St. Vincent.

In contrast, specific endemicity of lygaeids to the West Indies is high. One hundred and nineteen Antillean species do not occur on the mainland; thus 70% of the Antillean species occur only on the islands. Even if some relationships are masked by taxonomic provincialism or "oversplitting," a large proportion of the species occurs only on the islands.

The Lesser Antilles have only an occasional endemic species, and even these suggest inadequate collecting or taxonomic splitting. The situation is quite different in the Greater Antilles (Table 3-1). Of the 84 species known from Cuba, 23 (27%) are endemic there, and 48 are endemic species on one or the other of the four large islands of the Greater Antilles—28% of the entire West Indian fauna. This compares with only 6 species that are endemic to any or all of the remaining islands.

In order to try to understand what this means, we must look at the taxonomic units below the family level. I have used the subfamily units except for the large and diverse subfamily Rhyparochrominae, where tribal units seem more equivalent and yield more usable data. This method makes it apparent that speciation and hence endemicity vary enormously from taxon to taxon. For example, 35% of the endemism on Cuba is caused by the 8 species of Bledionotinae known to occur only there.

The distribution of the Bledionotinae in the West Indies is a good example of the dangers of using quantitative indices to establish relationships. The use of Simpson's faunal similarity index, whereby the total number of species on two islands is divided into the number of species the two islands have in common (× 100), shows that the faunas of Cuba and Jamaica have a higher similarity index than those of Cuba and Hispaniola (Table 3-2). These data are very misleading, primarily because of the presence of a number of species of Bledionotinae on both Cuba and Hispaniola, but *none* that are in com-

Table 3-2. Simpson's faunal similarity index as used in the distribution of Bledionotinae in the West Indies

Island	Genus	Species
Cuba-Jamaica	58	32.4
Cuba-Hispaniola	64	24.8
Cuba-Puerto Rico	66	32.6
Cuba-Dominica	54	25.5

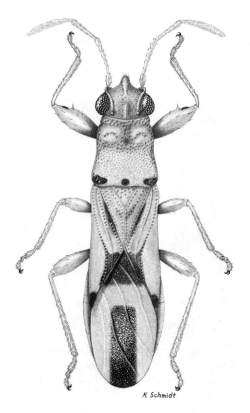

K. Schmidt

Figure 3-1. *Pamphantus trimaculatus* Slater, dorsal view. (From Slater 1981b, reprinted from the *Journal of the Kansas Entomological Society*, 54(1):85 (1981).

mon. Thus the bledionotine endemicity on Cuba and Hispaniola lowers the similarity index relative to Jamaica, where no Bledionotinae are known to occur, whereas actually the presence of the subfamily on the two islands indicates a closer relationship than either island has with Jamaica.

Perhaps no taxon better illustrates the complexities facing the lygaeidologist in attempting to understand the origins of the West Indian fauna than does the subfamily Bledionotinae (Fig. 3-1). Until the 1960s this subfamily, named Pamphantinae, was thought to be almost entirely restricted to the Greater Antilles. Scudder (1963) placed these species in the same subfamily as *Bledionotus*, a monotypic genus from the eastern Mediterranean. I subsequently described a striking genus, *Austropamphantus*, from northern Queensland (Slater 1981a). Such a distribution suggests a Gondwanaland origin for this subfamily with subsequent vicariance, but a word of caution is necessary. Most of the species are strikingly ant mimetic. Indeed, in the Antilles several form mimetic rings involving *Solenopsis* ants, other lygaeid bugs, spiders, and beetles (Myers and Salt 1926). This ant mimicry obscures some of the subfamilial recognition features. A careful analysis of relation-

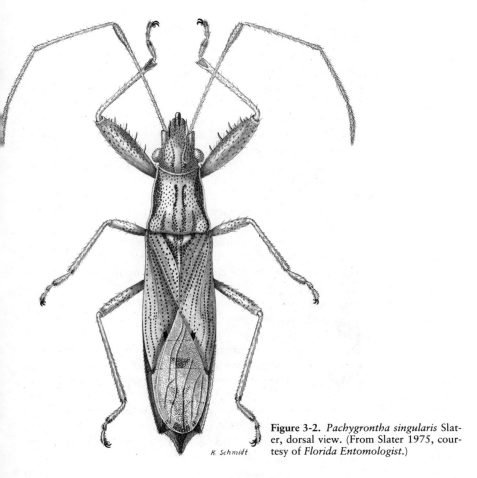

K. Schmidt

Figure 3-2. *Pachygrontha singularis* Slater, dorsal view. (From Slater 1975, courtesy of *Florida Entomologist.*)

ships is needed before we can be certain that we are dealing with a holophyletic group. If we are, it is one of the few taxa that shows close Old World relationships (besides a few semicosmopolitan taxa with, presumably, strong dispersal abilities).

The other taxon that may represent a Gondwanaland or West Gondwanaland element is *Pachygrontha singularis* (Fig. 3-2), which is known only from Cuba (Slater 1975). Although species of *Pachygrontha* occur in the tropics and subtropics of both hemispheres, this species appears to be phylogenetically more closely related to several West African species than to any of the South and Central American species.

In any event, endemism of the bledionotine species in the West Indies is 100%, and each of the species is restricted to a single island (Cuba, 7; Hispaniola, 5; Puerto Rico, 1).

The tribe Antillocorini also shows a high degree of endemicity. Two genera, *Bathydema* and *Antillodema*, are endemic and contain 73% of the known West Indian antillocorine species. I recently concluded that these endemic West Indian genera are closely related but not sister taxa (Slater 1980). Of the seven endemic species, six occur on only a single island (Jamaica, 3; Cuba, 2; Hispaniola, 1; St. Vincent, 1). This may be a conservative estimate, since another Cuban species has a slightly differentiated population on Puerto Rico. *B. sericea* occurs on Guadeloupe and Dominica and is the only species of these endemic genera found on more than one island.

The antillocorines are minute insects and are probably undercollected. Some if not all of the endemic species live on species of *Pilea*, often in montane areas. Several of the species from higher elevations are flightless. The occurrence of species through the Greater Antilles and again in the wet southern Lesser Antilles suggests that species may be (or were) present throughout the West Indies where available habitat exists. Their relationships are with South American taxa, but this is a group that has been on the islands a long time, as suggested by the speciation and by the occurrence of flightless morphs.

The second-highest degree of endemicity occurs in the rhyparochromine tribe Ozophorini. Only the single genus *Ozophora* is present on the islands. Distinct subgroups are present whose sister taxa are sometimes found in Mexico and Central America, and sometimes in South America. *Ozophora* species are abundant in the Neotropics and several occur in the Nearctic north of Mexico. Members of the genus may well occur on every island in the West Indies. However, 69% of the West Indian species do not occur on the mainland (if south Floridian species are included, the percentage rises to 89%). Of the 28 endemic species, 13 (46%) are known from a single island in the Greater Antilles (if the Caymans and Mona Island are included): Hispaniola, 5; Caymans, 3; Cuba, 2; Jamaica, 2; Mona, 1.

One of the most informative situations involves the "*quinquemaculata* complex." *O. rubrolinea* occurs from Mexico south through Central America and into South America, at least as far as Venezuela (Fig. 3-3). This species is readily recognizable by the presence of a pale narrow stripe down the middle of the dark macula mesally on the posterior pronotal lobe, a pale or often pink apical corial margin adjacent to the dark apical corial spot, and by details of the paramere shape and the relative length of the third antennal segment. A second species, *O. quinquemaculata*, occurs throughout most of the West Indies (Fig. 3-4). It is similar in appearance to *O. rubrolinea* but differs by its dark median pronotal area, dark apical corial margin, a shorter third antennal segment, and slight differences in paramere shape. Both of these species occur sympatrically on Jamaica and Hispaniola, and, at least on Jamaica, they can be taken in numbers at the same light trap on the same night. Whether they actually occur together in the field is unknown. Some

Kathy Schmidt

Figure 3-3. *Ozophora rubrolinea* Slater, dorsal view. (From Slater 1987, courtesy of *Florida Entomologist*.)

introgression is suggested on these islands, but the two appear to be quite distinct species. East of Hispaniola only *O. quinquemaculata* occurs. While the populations on Puerto Rico, Tortola, and some of the Virgins can be distinguished from those on Jamaica and Hispaniola by minor differences in antennal coloration, *O. quinquemaculata* populations do not change appreciably until one reaches Guadeloupe. From the latter island south to and including St. Lucia, the populations are almost identical in most respects with those to the north, but they differ in having upstanding hairs present on the dorsal surface of the body. This Guadeloupe to St. Lucia form is recognized as a distinct subspecies, *O. quinquemaculata subtilis* (Slater 1987). It occurs sympatrically with a species—*O. longirostris*—that has a very long beak and upstanding hairs on the dorsal surface; but it has the pronotal and corial coloration of *O. rubrolinea*, of which it is the sister species. Neither of these populations extend south of St. Lucia.

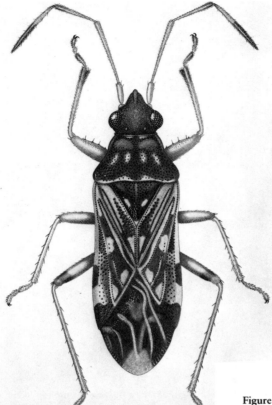

Kathy Schmidt

Figure 3-4. *Ozophora quinquemaculata* Barber, dorsal view. (From Slater 1987, courtesy of *Florida Entomologist*.)

On St. Vincent, Grenada, and Barbados a quite distinct species—*O. agilis*—occurs. It has a dark pronotal median macula, but the distal area of the apical corial margin adjacent to the dark apical corial spot is sexually dimorphic: pale in males, dark in females. It has a black or dark brown, rather than pale, first antennal segment. This species is found on Tobago, and it is the most abundant lygaeid at light traps in the northern range on Trinidad. It also appears to be widely distributed in northern South America.

What scenario can be visualized to account for such a situation (Fig. 3-5)? The status of *O. rubrolinea* and *O. agilis* seems clear. The former is primarily a Central American species that has reached the Greater Antilles overwater. Similarly *O. agilis* has all of the appearance of a South American species that has extended its range outward from the continental shelf onto the most southern of the offshore Windwards—Grenada and St. Vincent.

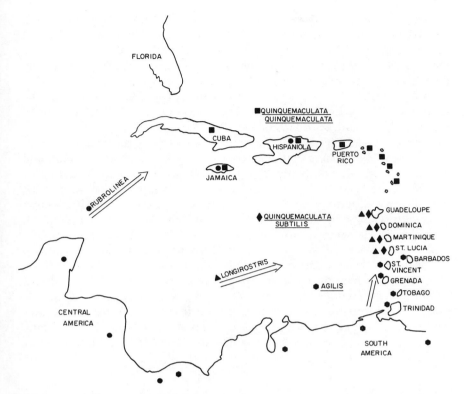

Figure 3-5. Distribution of members of the *Ozophora quinquemaculata* complex. (From Slater 1987, courtesy of *Florida Entomologist*.)

O. *quinquemaculata* has not been found on the mainland. Its sister species is probably an undescribed species from Mexico. My hypothesis is that its ancestral population reached the West Indies at an earlier period than did those of O. *rubrolinea* and O. *agilis* and differentiated there so that at least one of its populations has reached recognizable subspecific status. The occurrence of O. *quinquemaculata subtilis* and O. *longirostris* sympatrically on the "wet" islands from Guadeloupe to St. Lucia suggests that O. *q. subtilis* reached the area from the north as O. *quinquemaculata* spread eastward from an original entry in the western Greater Antilles, whereas O. *longirostris* represents the waif arrival of an ancestral O. *rubrolinea* stock. The fact that these two taxa resemble one another by the possession of short but distinct upstanding hairs on the dorsal body surface but differ markedly in beak length suggests introgression on the one hand and character displacement on the other. Comparative study of the ecology of O. *q. subtilis* and O. *longirostris* on these islands should be very rewarding.

I have discussed this complex in some detail because it appears to be a series of overwater dispersal events. Any other notion would support a belief in very long-term stasis in these populations, even though there is evidence that *Ozophora* is a genus whose members seem to be radiating extensively throughout the islands as well as Central and South America.

Only two other higher taxa show any degree of specific endemism on the islands—48% of Lygaeinae, and 28% of Myodochini (58% and 34%, respectively, if south Florida is included).

The tribe Myodochini is the dominant group of rhyparochromine bugs in the Western Hemisphere with a diverse and abundant South, Central, and North American fauna (Slater 1986). Although nine species do not occur outside of the West Indies, the endemism is not compelling because of the widespread distribution within the islands of most species. Only *Heraeus pulchellus* (Cuba and Grand Cayman), *Myodocha fulvosa* (Cuba), and *M. froeschneri* (Hispaniola) are restricted—essentially—to a single island. There are no West Indian endemics that do not occur in the Greater Antilles.

The degree of island endemism of Lygaeidae is generally predictable. Cuba and Hispaniola have the largest number of species restricted to them. Almost all of the single-island endemics are found only on the Greater Antilles. One hundred and thirty-seven species are known from the Greater Antilles (including the Caymans), of which 54 occur only there, a 39% endemism rate. By contrast, only three species are confined to a single island in the Lesser Antilles.

Wet Islands of the Southern Lesser Antilles

The southern islands of the Leewards and the Windward Islands are of particular importance because their topography provides a diversity of habitats, and because the proximity of some of the islands to the South American mainland provides stepping stones for the potential movement of elements of the South American fauna into the islands.

This discussion deals with the series of volcanic islands from Guadeloupe south through Grenada. These islands can be characterized as being wet on their eastern slopes and dry on the western. Sometimes the change is dramatic. There are, of course, frequent variations, with the southern portion being xeric and the northern portion rain forest. Martinique is a good example of this pattern.

There is a faunal break between St. Vincent and Grenada on the one hand and St. Lucia–Martinique–Dominica–Guadeloupe on the other. Grenada is only 90 miles (146 kilometers) from Tobago and is the closest island to the continental shelf. Between Grenada and St. Vincent lies a shallow sea interrupted by the numerous small islands of the Grenadines. St. Vincent and

Grenada may well have been connected during interglacial maxima. In any event, the two islands have several South American species that do not extend north of them. Examples are *Ozophora agilis, Myodocha unispinosa, Prytanes oblonga, P. micropterum* and *Erlacda gracilis.* A few species that are otherwise South American are also found only on Grenada. *Blissus planus* is apparently endemic to Grenada and *Bathydema socia* is known only from the volcano on St. Vincent.

In addition to having a South American faunal element that apparently reaches no further north, these southern islands themselves appear to have served as steppingstones for additional species that occur on them and extend northward, often to—but not north of—Guadeloupe. Such species are *Prytanes formosa, Botocudo picturata, Cryphula affinis, Bubaces uhleri* (?), *Valtissius distinctus, Geocoris lividipennis,* and a subspecies of *Pachygrontha minarum* (ssp. *saileri*).

A sea level as much as 600 feet (472 meters) lower than at present would not only effect a land connection between Grenada and St. Vincent but would dramatically increase the size of the Lesser Antillean islands to the north. Thus, islands such as Martinique, Dominica, Guadeloupe, and St. Lucia, although not connected by land with St. Vincent, would be in much closer proximity to one another (they are not very far apart at present), greatly facilitating faunal interchange. This view is supported by the overall similarity of the faunas of these islands and the almost complete lack of endemicity on any of them. It is unfortunate that our data from the smaller islands to the north are too scanty to even begin to interpret the effect of large size on these islands and their relationships to a large Bahaman plateau.

A real faunal break between St. Lucia and St. Vincent is evidenced not only by the South American elements that do not extend northward of St. Vincent but also by the previously discussed situation in the genus *Ozophora* and a number of additional species of *Ozophora* that do not extend south of St. Lucia.

West Indies—Florida Relationships

It is useful to compare the West Indian species that also occur in Florida. There are 48 such species (29% of the West Indian fauna) (Table 3-3). Of these, 23 are widespread on the mainland, most occurring in the southern United States through Central America into at least northern South America. They tend to be species of oligophagous feeding habits, the "weed species," and tend to be widespread on the islands.

According to Bond (1978), the greater part of the West Indian passerine bird fauna is North American rather than South American in origin. This is primarily a Central American tropical element, however. Baker and Geno-

Table 3-3. Lygaeidae: Florida and West Indies

	Florida genera	Florida species	Genera in common with West Indies	Species in common with West Indies
Lygaeinae	6	14	5	9
Orsillinae	3	7	3	5
Ischnorhynchinae	1	2	1	1
Cyminae	3	6	2	2
Blissinae	2	12	2	3
Geocorinae	2	5	1	2
Pachygronthinae	2	5	1	2
Lethaeini	3	3	3	1
Antillocorini	3	5	2	2
Drymini	1	1	0	0
Ozophorini	1	7	1	6
Myodochini	15	26	8	15
Rhyparochromini	1	1	0	0
Gonianotini	3	3	0	0
Total	46	97	29 (63%)	48 (49%)

ways (1978), in an analysis of the West Indian bat fauna, believe that Florida served as a minor but probable source area and that the Floridian bat fauna is not Caribbean in nature. Since bats and lygaeids both fly, does this correspond with evidence from the Lygaeidae? I suggest that whereas there are a few lygaeids that have obviously reached the West Indies from Florida, the reverse is much more significant and that a considerable proportion of the south Floridian fauna has been derived from the West Indies. Moreover, this process is still going on both directly from Cuba and indirectly via the Bahama Islands stepping stones. It appears that of the 48 West Indian–Floridian species only 9 can be considered to have reached the West Indies via Florida, whereas approximately 24 appear to have reached Florida from the islands.

A number of West Indian species that also occur in the United States are restricted to south Florida. Most, if not all, appear to be there as a result of overwater dispersal. Some, such as *Ozophora levis*, *O. reperta*, and *Prytanes dissimilis*, probably have entered from the Greater Antilles via the Bahamas. In other cases they have reached Florida directly from one or the other of the Greater Antilles—probably most often Cuba. Good examples of the latter are *Heraeus triguttatus* and *Oedancala cubana*. *O. cubana* is the only grass-feeding species of the genus and is known only from Big Pine Key in Florida, and Cuba, suggesting a recent direct introduction. *P. intercisa*, which occurs in other Gulf states, may represent a similar phenomenon. There probably are other cases we do not know about. For example, *O. caroli*, thought to be a south Florida endemic, is now known from Hispaniola.

Summary

Knowledge of the lygaeid fauna of the West Indies is to a considerable extent still in the data-gathering stage. There is a high degree of specific endemicity but very low generic endemicity. Species that are confined to single islands are almost entirely in the Greater Antilles, which presumably reflects not only the size and diversity but also the relative antiquity of the Greater Antilles.

Although most taxa have not been subjected to rigorous cladistic analysis, present evidence suggests that most of the fauna has accumulated from mainland source areas by overwater dispersal. Only in the Bledionotinae, in one species of *Pachygrontha*, and possibly in the endemic antillocorine genera does the possibility exist that the fauna represent mainland-island vicariance.

The belief that congruence of distribution patterns of different groups implies vicariance may not be applicable to islands lying relatively close to different source areas.

Although general patterns of distribution of West Indian Lygaeidae substantially agree with those of many other groups, careful collecting on the smaller islands and careful cladistic analysis suggest that the family can be used successfully for a better biogeographic understanding of this complex area.

Acknowledgments

I wish to extend my appreciation to the following for fruitful discussions of West Indian biogeography and for bringing pertinent literature to my attention: R. M. Baranowski at the University of Florida, Homestead; and George Clark, Jane Ellen O'Donnell, Steve Pacala, David Taylor, and Rick Wetzler at the University of Connecticut.

I thank Peter Ashlock of the University of Kansas for allowing me to use one of his manuscript names; Mary Jane Spring of the University of Connecticut and Kathleen Schmidt, formerly of the American Museum of Natural History for preparation of the illustrations; and Jane O'Donnell and Elizabeth Slater for extensive assistance in the preparation of the manuscript.

This work was supported in part by a grant from the National Science Foundation.

Appendix 3.1: Synopsis of Higher Group Taxa in the West Indies (see Table 3-4)

Lygaeinae: The fauna is relatively diverse and includes (in the West Indies) the second largest number of genera of any major taxon. There is considerable specific endemicity with single-island endemics found chiefly in the genus *Ochrimnus*. Five of the seven genera present contain endemic species. Of the fifteen species known only from the West Indies, eleven occur only on a single island.

Orsillinae: The fauna is depauperate and confined largely to widespread species of oligophagous feeding habits living in early succession, "weed" habitats. There is a definite North American element illustrated by such widespread species as *Xyonysius californicus* and *Nysius raphanus*. No endemicity is present on the islands.

Ischnorhynchinae: The fauna is depauperate, consisting only of two species of *Kleidocerys*, the widespread Neotropical *virescens* and a Cuban endemic, *suffusus*.

Cyminae: The fauna is depauperate, consisting only of three widespread Neotropical species.

Blissinae: Five of the nine known species are confined to a single island. Of these, *Patritius cubensis* and *Toonglasa discalis* occur only on Cuba, and at least the former may represent a vicariant element. The endemic species of *Blissus* are recently described and need further study. They include *B. planus* from Grenada, one of the very few single-island endemics from the Lesser Antilles. *Blissus* is an extremely difficult taxon and species limits are often obscure.

Geocorinae: The fauna is relatively depauperate. Three elements are present on the islands: an old Neotropical genus, *Ninyas*, with an endemic species on Cuba (possibly additional undescribed species in the Greater Antilles); a South and Central American species of *Geocoris*; and a recent invasive North American element typified by such widespread species as *G. uliginosus* and *G. punctipes*.

Bledionotinae: All species and two of the three genera are endemic to the Greater Antilles. This subfamily may represent a vicariant element, as discussed in detail above.

Pachygronthinae: Two genera are present, including a remarkable Cuban endemic species of *Pachygrontha* which is apparently related to West African taxa. In *Pachygrontha* one species in the Greater Antilles, *P. longiceps*, is conspecific with Central American populations, contrasting with a subspecies of a South American species, *P. minarum*, found in the southern Lesser Antilles. Several species of both genera are widespread. *Oedancala cubana* is morphologically the most isolated member of the genus and is known only from Cuba and Big Pine Key, Florida.

Rhyparochrominae: This is the largest and most diverse of any lygaeid subfamily. It is analyzed here by tribe. The recent analysis of the world fauna by Slater (1986) concluded that the Neotropical fauna is primarily composed of species belonging to four tribes: Lethaeini, Antillocorini, Ozophorini, and Myodochini. Only one West Indian species does not belong to one of these four tribes.

Lethaeini: The fauna is surprisingly depauperate on the islands, with little or no endemicity, although it is diverse and abundant on the Neotropical mainland.

Antillocorini: Two endemic genera are present and there is a high degree of specific endemicity. There has been speciation of *Bathydema* in the Greater Antilles. Flightless morphs of several species of this genus occur in the mountains. These probably represent old faunal elements (Slater 1977). The antillocorines are very small, often cryptic insects, and probably more species remain to be discovered in this than in any other lygaeid group.

Ozophorini: A single genus is recognized at present on the islands. There has been extensive interisland radiation. Several species complexes are present. The fauna suggests several dispersal events from both Central and South America.

Myodochini: The tribe is numerous both in species and individuals. Many species occur in ruderal habitats and are widespread throughout the islands. Few species are endemic to the West Indies, and only two are restricted to a single island. This tribe is the dominant rhyparochromine element in the Western Hemisphere (Slater 1986).

Udeocorini: Only a single species occurs in the West Indies. It is widespread in the Western Hemisphere, even reaching Hawaii. The tribe is essentially Neotropical and Australian.

Table 3-4. West Indian Lygaeid fauna

	Number of species	% Species	Number of genera	% Genera
Ozophorini	36	21	1	2
Myodochini	32	19	9	18
Lygaeinae	31	18	8	16
Bledionotinae	13	8	3	6
Antillocorini	11	6.5	5	10
Blissinae	10	6	4	8
Geocorinae	9	5	2	4
Lethaeini	7	4	6	12
Orsillinae	7	4	3	6
Pachygronthinae	7	4	2	4
Cyminae	3	2	3	6
Ischnorhynchinae	2	1	1	2
Udeocorini	1	—	1	2
Cleradini	1	—	1	2
Total	169/170		49	

Appendix 3.2: Checklist of West Indian Lygaeidae (Species in *boldface* are endemic in the West Indies)

Lygaeinae
1. *Oncopeltus semilimbatus* Stal
2. *Oncopeltus aulicus* (F.)
3. **Oncopeltus cingulifer** ssp. **antillensis** O'Rourke
4. *Oncopeltus sandarachatus* (Say)
5. *Oncopeltus fasciatus* (Dallas)
6. **Oncopeltus spectabilis** Van Duzee, Jamaica
7. *Oncopeltus cayensis* Torre Bueno
8. **Oncopeltus pictus** Van Duzee
9. *Oncopeltus sexmaculatus* Stal (?)
10. **Lygaeus wygodzinskyi** Alayo, Cuba
11. *Lygaeus bahamensis* Barber and Ashlock
12. *Lygaeus formosus* Blanchard
13. **Lygaeus coccineus** Barber
14. *Lygaeus alboornatus* (Blanchard) (introduced)
15. **Lygaeus dearmasi** Alayo, Cuba
16. *Lygaeus kalmii* Stal

17. *Ochrimnus collaris* (F.)
18. *Ochrimnus dallasi* (Guerin)
19. *Ochrimnus henryi* Brailovsky, Puerto Rico
20. *Ochrimnus laevus* Brailovsky, St. Kitts
21. *Ochrimnus nigriceps* (Scudder), Grand Cayman
22. *Ochrimnus pseudocollaris* Brailovsky
23. *Ochrimnus tripligatus* (Barber)
24. *Craspeduchus bilimeki* (Distant)
25. *Craspeduchus pulchellus* (F.)
26. *Torvochrimnus confraternus* (Uhler)
27. *Torvochrimnus poeyi* (Guerin), Cuba
28. *Melanopleurus maculicorium* Maldonado-Capriles, Hispaniola
29. *Melanopleurus tetraspilus* Stal, Cuba
30. *Neacoryphus albonotatus* (Barber)
31. *Neacoryphus nigrinervis* (Stal)

Orsillinae
32. *Neortholomus jamaicensis* (Dallas)
33. *Neortholomus koreshanus* (Van Duzee)
34. *Xyonysius californicus* (Stal)
35. *Xyonysius basalis* (Dallas)
36. *Nysius tenellus* Barber
37. *Nysius raphanus* Howard
38. *Nysius scutellatus* Dallas

Ischnorhynchinae
39. *Kleidocerys suffusus* Barber, Cuba
40. *Kleidocerys virescens* (F.)

Cyminae
41. *Cymodema breviceps* (Stal)
42. *Cymoninus notabilis* (Distant)
43. *Neoninus illustris* Distant

Blissinae
44. *Patritius cubensis* Barber, Cuba
45. *Blissus insularis* Barber
46. *Blissus slateri* Leonard, Puerto Rico
47. *Blissus planus* Leonard, Grenada
48. *Blissus antillus* Leonard, Puerto Rico
49. *Toonglasa discalis* (Barber), Cuba
50. *Ischnodemus fulvipes* (De Geer)
51. *Ischnodemus praecultus* Distant
52. *Ischnodemus* sp.

Geocorinae
53. *Ninyas deficiens* (Lethierry)
54. *Ninyas humeralis* Barber, Cuba

55. *Ninyas strabo* Distant
56. **Geocoris "disparatus"** Johnson and Fox (ms), Jamaica
57. *Geocoris lividipennis* Stal
58. *Geocoris punctipes* (Say)
59. *Geocoris thoracicus* (Fieber)
60. *Geocoris pallens* Stal (?)
61. *Geocoris uliginosus* (Say)

Bledionotinae
62. **Abpamphantus gibbosus** Barber, Cuba
63. **Neopamphantus calvinoi** Barber and Bruner, Cuba
64. **Neopamphantus maculatus** Barber and Bruner, Hispaniola
65. **Neopamphantus hispaniolus** Slater, Hispaniola
67. **Pamphantus atrohumeralis** Barber and Bruner, Hispaniola
    Pamphantus atrohumeralis ssp. **dominicanus** Slater, Hispaniola
68. **Pamphantus elegantulus** Stal, Cuba (+ I. Pines)
69. **Pamphantus mimeticus** Barber, Cuba
70. **Pamphantus pallidoides** Slater, Hispaniola
71. **Pamphantus pallidus** Barber and Bruner, Cuba
72. **Pamphantus pellucidus** Slater, Puerto Rico
73. **Pamphantus stenoides** (Guerin) (?), Cuba
74. **Pamphantus trimaculatus** Slater, Hispaniola
75. **Pamphantus vittatus** Bruner, Cuba

Pachygronthinae
76. *Oedancala bimaculata* (Distant)
77. *Oedancala crassimana* (F.)
78. *Oedancala cubana* Stal
79. *Pachygrontha compacta* Distant
80. *Pachygrontha longiceps* Stal
81. **Pachygrontha minarum** ssp. **saileri** Slater
82. **Pachygrontha singularis** Slater, Cuba

Rhyparochrominae
Lethaeini
83. *Cistalia signoreti* (Guerin)
84. *Cryphula affinis* (Distant)
85. *Cryphula fasciata* (Distant) (?)
86. *Paragonatas divergens* (Distant)
87. **Valtissius distinctus** (Distant)
88. **Bubaces uhleri** (Distant)
89. N. g. N. sp.

Antillocorini
90. *Cligenes distinctus* Distant
91. **Bathydema darlingtoni** Slater and Baranowski, Jamaica
92. **Bathydema cubana** Slater and Baranowski, Cuba
93. **Bathydema hispaniola** Slater and Baranowski, Hispaniola

94. *Bathydema socia* Uhler, St. Vincent
95. *Bathydema jamaicensis* Slater and Baranowski, Jamaica
96. *Bathydema sericea* (Lethierry)
97. *Antillodema obscura* (Barber), Cuba
98. *Antillodema maculosa* Slater and Baranowski, Jamaica
99. *Antillocoris pallidus* (Uhler)
100. *Botocudo picturatus* (Distant)

Cleradini
101. *Clerada apicicornis* Signoret (introduced)

Ozophorini
102. *Ozophora atropicta* Barber
103. *Ozophora burmeisteri* (Guerin)
104. *Ozophora cubensis* Barber, Cuba
105. *Ozophora divaricata* Barber
106. *Ozophora pallidifemur* Scudder, Caymans
     Ozophora pallidifemus ssp. *fuscifemur* Scudder, Caymans
107. *Ozophora reperta* Blatchley
108. *Ozophora miniscula* Scudder, Caymans
109. *Ozophora octomaculata* Ramos, Mona
110. *Ozophora longirostris* Slater and Baranowski
111. *Ozophora josephina* Slater and Baranowski
112. *Ozophora pallescens* (Distant)
113. *Ozophora quinquemaculata* Barber
     Ozophora quinquemaculata ssp. *subtilis* Slater
114. *Ozophora hirsuta* Slater and Baranowski
115. *Ozophora subimpicta* Barber
116. *Ozophora majas* Baranowski and Slater
117. *Ozophora caribbae* Baranowski and Slater
118. *Ozophora helenae* Baranowski and Slater
119. *Ozophora cobbeni* Scudder
120. *Ozophora hispaniola* (ms), Hispaniola
121. *Ozophora testacea* (ms), Hispaniola
122. *Ozophora laticephala* Slater and O'Donnell
123. *Ozophora darlingtoni* (ms), Hispaniola
124. *Ozophora alayoi* (ms), Cuba
125. *Ozophora caliginosus* (ms), Hispaniola
126. *Ozophora pusilla* (ms)
127. *Ozophora caroli* Slater and Baranowski
128. *Ozophora coleoptrata* (ms)
129. *Ozophora archboldi* (ms)
130. *Ozophora levis* Slater and Baranowski
131. *Ozophora hohenbergia* Slater and Baranowski, Jamaica
132. *Ozophora nitidus* Slater, Jamaica
133. *Ozophora rubrolinea* Slater
134. *Ozophora umbrosa* (ms)
135. *Ozophora agilis* Slater

Myodochini
136. *Prytanes confusa* (Barber)
137. *Prytanes dissimilis* (Barber)
138. *Prytanes intercisa* (Barber)
139. *Prytanes minima* (Guerin)
140. *Prytanes oblonga* (Stal)
141. *Prytanes formosa* (Distant)
142. **Prytanes cubensis** (Barber, Cuba
143. *Prytanes micropterum* Scudder (ms)
144. *Heraeus guttatus* (Dallas)
145. *Heraeus plebejus* Stal
146. **Heraeus pulchellus** Barber
147. *Heraeus triguttatus* (Guerin)
148. *Heraeus* sp.
149. *Neopamera albocincta* (Barber)-complex
150. **Neopamera albocincta-complex**
151. **Neopamera albocincta-complex**
152. **Neopamera albocincta-complex**
153. *Neopamera bilobata* (Say)
     Neopamera bilobata ssp. *scutellata* (Dallas)
154. **Neopamera intermedia** (Barber)
155. *Neopamera neotropicalis* (Kirkaldy)
156. **Neopamera vicarius** (Barber)
157. *Neopamera vicinalis* (Distant)
158. *Pseudopachybrachius vinctus* (Say)
159. *Pseudopachybrachius basalis* (Dallas)
160. *Pseudopachybrachius concepcioni* Zheng and Slater
161. *Paromius longulus* (Dallas)
162. **Paromius dohrnii** (Guerin)
163. **Myodocha fulvosa** Barber, Cuba
164. *Myodocha unispinosa* Stal
165. **Myodocha froeschneri** (ms), Hispaniola
166. *Froeschneria multispinus* (Stal)
167. *Froeschneria piligerus* (Stal)
168. *Ligyrocoris litigiosus* (Stal)
169. *Erlacda gracilis* (Uhler)

Udeocorini
170. *Tempyra biguttula* Stal

References

Baker, R.J., and H.H. Genoways. 1978. Zoogeography of Antillean Bats. *In* Zoogeography in the Caribbean, pp. 53–97. Special Publ. No. 13, Acad. Nat. Sci. Philadelphia.
Baranowski, R.M., and J.A. Slater. 1975. The life history of *Craspeduchus pulchel-*

lus, a lygaeid new to the United States (Hemiptera: Lygaeidae). Florida Entomol. 58:297–302.

Bond, J. 1978. Derivations and continental affinities of Antillean Birds. *In* Zoogeography in the Caribbean, pp. 119–128. Special Publ. No. 13, Acad. Nat. Sci. Philadelphia.

Brown, F.M. 1978. The origins of the West Indian butterfly fauna. *In* Zoogeography in the Caribbean, pp. 5–30. Special Publ. No. 13, Acad. Nat. Sci. Philadelphia.

Coney, P.J. 1982. Plate tectonic constraints on the biogeography of Middle America and the Caribbean Region. Ann. Missouri Bot. Garden 69:432–443.

Matile, L. 1982. Systematique, phylogenie et biogéographie des Dipteres Keroplatidae des Petites Antilles et de Trinidad. Bull. Mus. Nat. Hist. Paris 4:ser. 4:sect. A (1–2):189–235.

Myers, J.G., and G. Salt. 1926. The phenomenon of myrmecoidy with new examples from Cuba. Trans. Entomol. Soc. London 1926:427–436.

Popov, Y.A. 1981. Historical development and some questions on the general classification of Hemiptera. Rostria Supp. 33:85–99.

Rosen, D.E. 1985. Geological hierarchies and biogeographic congruence in the Caribbean. Ann. Missouri Bot. Garden 71(4):636–659.

Scudder, G.G.E. 1963. Pamphantinae, Bledionotinae and the genus *Cattarus* Stal (Hemiptera: Lygaeidae). Opusc. Entomol. 28:81–89.

Slater, J.A. 1975. The Pachygronthinae of the West Indies, with the description of a new species of Pachygrontha from Cuba (Hemiptera: Lygaeidae) Fla. Entomol. 58:65–74.

———. 1977. The incidence and evolutionary significance of wing polymorphism in Lygaeid bugs with particular reference to those of South Africa. Biotropica 9:4:217–229.

———. 1980. Systematic relationships of the Antillocorini of the Western Hemisphere (Hemiptera: Lygaeidae). Syst. Entomol. 5:199–226.

———. 1981a. Two new genera of Lygaeidae from Northern Australia including the first member of the Pamphantini from the Eastern Hemisphere (Hemiptera: Heteroptera). J. Australian Entomol. Soc. 20:111–118.

———. 1981b. New species of Pamphantini from South America and the West Indies (Hemiptera: Lygaeidae). J. Kansas Entomol. Soc. 54:83–94.

———. 1986. A synopsis of the zoogeography of the Rhyparochrominae (Hemiptera: Lygaeidae). J. New York Entomol. Soc. 94:(2):262–280.

———. 1987. The systematic status of *Ozophora quinquemaculata* Barber and related taxa (Hemiptera: Lygaeidae). Florida Entomol. 70:134–162.

Slater, J.A., and J.E. O'Donnell. 1979. An analysis of the *Ozophora laticephala*-complex with the description of eight new species (Hemiptera: Lygaeidae). J. Kansas Entomol. Soc. 52:154–179.

Wilson, E.O. 1985. Ants of the Dominican amber (Hymenoptera: Formicidae), part 2: The first fossil army ants. Psyche 92:11–16.

4 · Zoogeography of the Auchenorrhynchous Homoptera of the Greater Antilles (Hemiptera)

J. A. Ramos

This paper presents a zoogeographic analysis of the auchenorrhynchous fauna of the Greater Antilles. The Greater Antilles consist of the four major West Indian islands: Cuba, the largest with an area of 114,000 km², is basically a flat plain stretching east to west, with rather high mountains on both ends. Hispaniola, second largest in area (77,250 km²) and more complex physiographically, has high mountains running from east to west in both the north and south. Its Pico Duarte, the highest peak in the West Indies, is over 3000 m high. Jamaica and Puerto Rico are similar to each other in some aspects although Jamaica (10,896 km²) is about one-fourth larger than Puerto Rico (8,676 km²). Both are very mountainous centrally with narrow coastal plains. They are not as diverse as their larger sisters.

The Auchenorrhyncha is a suborder of the Homoptera, which includes the insects commonly known as cicadas (Cicadoidea), treehoppers (Membracoidea), spittlebugs or froghoppers (Cercopoidea), planthoppers (Fulgoroidea), and leafhoppers (Cicadelloidea). According to Metcalf (1947), the Homoptera serve well as zoogeographic indicators because many are relatively small in size, not highly mobile, and, being wholly phytophagous, are generally restricted to a single plant host or to a narrow range of host plants. In addition, their dispersion is apparently influenced by other ecological factors not yet understood.

Puerto Rico is the only of the Greater Antilles whose auchenorrhynchous fauna has been thoroughly studied taxonomically. Caldwell (1952), Caldwell and Martorell (1952), and Ramos (1957) revised all of the families represented on the island. Cuba is fairly well known, as Metcalf and Bruner (1925a, 1925b, 1930, 1936, 1944, 1948, and 1949) monographed the Membracidae, Cercopidae, Cicadellidae, and several families of the Fulgoroidea. Dlabola (1977) gave the number of the species in the auchenorrhynchous families for Cuba and Puerto Rico. Hispaniola is poorly known with the exception of the Membracidae and Cicadidae, which Ramos (1979, 1983) recently revised.

Jamaica has the least known auchenorrhynchous fauna in the Greater Antilles. Van Duzee (1907) gave valuable information on the distribution of Jamaican species, but no modern revision or survey of the fauna has been done.

Young's (1977) extraordinary work on the taxonomy of the Cicadellinae of the New World permitted an analysis of the leafhopper fauna from the Greater Antilles, and Metcalf's *General Catalogue of the Homoptera* (1945) provided valuable information on the distribution of genera and species.

Cicadoidea

Twenty-eight species of cicadas, placed among 10 genera and 2 families (Table 4-1), are known from the Greater Antilles. Of these, 24 species are exclusively Greater Antillean, representing an endemism level of 86% (Table 4-2). The 4 species of nonendemics are widespread taxa in North, Central, and South America. The percentage of endemism for the entire group of islands seems outstanding as compared to other groups of insects. Brown (1978) recorded endemism of 54% for the butterflies of all the West Indies. Wolda (1984) indicated that Panama, with 30 species of cicadas, has a rich fauna. The number of cicadas recorded from the islands justifies a similar appreciation for the Greater Antilles.

Of the 4 genera of cicadas endemic to the Greater Antilles (Table 4-1), *Juanaria* is restricted to Cuba, *Psallodia* to Hispaniola, *Borencona* to Puerto

Table 4-1. Distribution of Greater Antillean genera of Cicadoidea

Genus	North America	Mexico, Central America	South America	Cuba	Hispaniola	Jamaica	Puerto Rico
Family Cicadidae							
Odopoea		+	+		+	+	
Juanaria				+			
Borencona							+
Chinaria		+			+		
Uhleroides				+	+		
Diceroprocta	+	+	+	+		+	
Proarna	+	+	+	+	+	+	+
Ollanta		+			+		
Family Tibicinidae							
Psallodia					+		
Cicadetta	+				+		

Table 4-2. Species level endemism of Greater Antillean Cicadoidea

Island	No. of species	No. of endemics	Percentage of endemism
Cuba	10	7	70
Hispaniola	13	11	85
Jamaica	7	5	71
Puerto Rico	2	1	50
Greater Antilles	28	24	86

Rico, and *Uhleroides* is shared by Cuba and Hispaniola. *Juanaria* and *Borencona* have affinities with *Chinaria*, also known from Mexico; and *Uhleroides* with *Odopoea*, which occurs in Hispaniola, Jamaica, and from Mexico to Argentina. Of 14 species known in *Odopoea*, 8 (57%) occur in Hispaniola and Jamaica. This interesting distribution illustrates a great degree of vicariance in *Odopoea* in these islands. *Psallodia*, the only endemic tibicinid in the Greater Antilles, is allied to *Taphura*, which ranges from Panama to Brazil. *Ollanta* is known also from Mexico and South Caicos Island. The remaining genera, *Diceroprocta*, *Proarna*, and *Cicadetta*, are widely distributed in the New World.

The presence in Hispaniola of one endemic species of cicada in each of the genera *Chinaria* and *Ollanta* could be very significant. Because both genera possibly originated in Mexico, their presence in the Greater Antilles indicates an affinity between the Cicadidae of the islands and the fauna of Mexico and Central America.

Membracidae

The membracids are represented in the Greater Antilles by 64 species grouped into 26 genera and 4 subfamilies (Table 4-3). The Centrotinae are believed to represent the most primitive group in the Membracidae. The vast majority of the members of the Centrotinae are of African and Oriental distribution. In the New World the Centrotinae are represented by a total of 17 genera, of which 10 occur in Mexico, 5 in Central America, 4 in the Greater Antilles, 2 in North America, and 2 in South America. This again illustrates a strong affinity between the Centrotinae of the Greater Antilles and the Mexican and Central American fauna, but little affinity between the Centrotinae of the islands and the North and South American fauna.

The remaining three subfamilies of the Membracidae represented in the Greater Antilles are the Membracinae, Smiliinae, and Nessorhininae. The first two of these subfamilies are rather poorly represented in the islands, both in the number of genera and of endemics. Of 33 genera described in the

Table 4-3. Distribution of Greater Antillean genera of Membracidae

Genus	North America	Mexico, Central America	South America	Cuba	Hispaniola	Jamaica	Puerto Rico
Brachycentrotus				+	+		+
Daimon					+		
Orthobelus				+	+	+	
Monobelus				+	+	+	+
Callicentrus					+	+	
Paradarnoides*							+
Spathenotus							+
Spinodarnoides			+				+
Goniolomus				+			
Nessorhinus				+	+	+	
Orekthophora					+		
Marshallella					+	+	
Brachytalis				+			+
Monobeloides					+		
Bolbonota	+	+	+				+
Jibarita							+
Umbonia	+	+	+				+
Platycotis	+	+	+	+	+		
Stalotypa				+			
Micrutalis	+	+	+	+	+	+	+
Spissistilus	+	+		+	+		+
Vanduzeea	+	+	+		+		
Idioderma	+			+		+	
Antianthe	+	+	+		+		
Phormophora			+			+	
Quadrinarea						+	

*Also known from the Lesser Antilles.

Membracinae, 26 (79%) occur in South America. In contrast, only 5 genera (15%) are known from the Greater Antilles (2 endemics—*Stalotypa* from Cuba and *Jibarita* from Puerto Rico). The tribe Membracini is represented only in Puerto Rico with the genera *Bolbonota* and *Jibarita*. The first is widely distributed throughout South America. *Jibarita* is a Puerto Rican genus allied to *Hypsoprora*, present in Central and South America.

The Similiinae have 88 known genera of which 5, or 6%, are represented in the Greater Antilles. The smiliine tribe Quadrinareini, with only one genus (*Quadrinarea*) is endemic to Jamaica and represents a unique and striking feature of the membracid fauna of the island.

The Nessorhininae are apparently related to the Centrotinae (Deitz 1975) and possibly may have been derived from them. Of the 26 recorded genera, 10 belong to the Nessorhininae. This constitutes 91% of all the genera placed under that subfamily (Deitz 1975, Ramos 1979), making it largely Greater Antillean in character. Hispaniola has 3 endemic genera, *Daimon*,

Table 4-4. Species level endemism of Greater Antillean Membracidae

Island	No. of species	No. of endemics	Percentage of endemism
Cuba	22	14	63
Hispaniola	31	21	67
Jamaica	10	6	60
Puerto Rico	16	9	56
Greater Antilles	64	49	76

Orekthophora, and *Monobeloides*; Cuba has 2, *Goniolomus* and *Stalotypa*; Puerto Rico has 2, *Spathenotus* and *Jibarita*; and Jamaica has one, *Quadrinarea*. Surprisingly, the Darninae, Heteronotinae, and Stegaspidinae, which are of wide distribution in North, Central, and South America, have not been recorded from the Greater Antilles.

A total of 49 membracid species and 14 genera are indigenous to the islands. This represents an endemism of 76% at the species level (Table 4-4) and of 54% at the generic level. These numbers indicate a rich and highly endemic membracid fauna.

Cercopoidea

Of the Greater Antilles only the Cercopoidea from Cuba and Puerto Rico have been worked faunistically (Metcalf and Bruner 1944; Ramos 1957). Information about the cercopoid fauna of Hispaniola and Jamaica is so scant and fragmentary that the following analysis will concentrate mainly on Cuba and Puerto Rico.

Three of the 4 families grouped into the Cercopoidea are represented in the Greater Antilles, except in Puerto Rico, where only the Aphrophoridae and Clastopteridae occur. Of the Cercopidae, the Tomaspidinae, which is restricted to the New World, is represented in Cuba with 3 species. One of these, *Prosapia bicincta* (Say), also recorded from Hispaniola and Jamaica, is a common species in the southern United States (Hamilton 1977). A second Cuban species is an endemic member of the genus *Neaenus*, which is also known from Mexico. The third Cuban tomaspidine is an endemic of uncertain generic position (Table 4-5).

The Aphrophoridae constitute the richest family of the Cercopoidea, both in the number of genera and species and in endemics. Of 9 genera known from the islands, 7 (78%) are endemic and 2 are widely distributed. The third family, the Clastopteridae, includes a large group of species of a single genus with wide distribution in the New World.

A total of 34 species of cercopoids are known from the Greater Antilles as

Table 4-5. Distribution of Greater Antillean genera of Cercopoidea

Genus	North America	Mexico, Central America	South America	Cuba	Hispaniola	Jamaica	Puerto Rico
Family Cercopidae							
Prosapia	+	+		+	+	+	
Neaenus				+			
Uncertain				+			
Family Aphrophoridae							
Enocomia				+	+	+	
Cephisus	+	+	+	+			
Lepyronia	+	+	+	+			
Leocomiopsis				+			+
Leocomia				+			+
Asprocranites							+
Gaeta				+			
Gaetopsis							+
Dasyoptera				+			
Family Clastopteridae							
Clastoptera	+	+	+	+	+	+	+

Table 4-6. Species level endemism of Greater Antillean Cercopoidea

Island	No. of species	No. of endemics	Percentage of endemism
Cuba	22	18	82
Hispaniola	3	2	66
Jamaica	3	2	66
Puerto Rico	8	5	62
Greater Antilles	34	27	79

follows: 22 in Cuba (18 endemic, or 82%); 8 in Puerto Rico (5 endemic, or 62%); and 2 each in Hispaniola and Jamaica. Twenty-seven species are restricted to the Greater Antilles, which is equivalent to 79% endemism for the islands as a group (Table 4-6).

Fulgoroidea

The superfamily Fulgoroidea constitutes a large, complex group of families little known in the Greater Antilles, except in Puerto Rico (Caldwell 1952, Caldwell and Martorell 1952, Ramos 1957) and in Cuba, where several families were studied by Metcalf and Bruner (1930, 1948). Of the 18 families into which the group may be divided (O'Brien and Wilson 1985), 11 are known from the Greater Antilles.

For convenience and brevity, the Kinnaridae was selected to represent the

Fulgoroidea in this analysis. The family is a small one, well studied by Fennah (1942, 1945, 1947, 1948) and Ramos (1957). Its center of diversity seems to occur in the Greater Antilles, with an endemism of 78% in the level of genera and 100% in the species level. Of 9 genera (Table 4-7) known from the Greater Antilles, 7 are endemic. The total number of genera catalogued for the family is 13. Of this number, 2 are extralimital and 2 occur only in the Lesser Antilles.

The level of endemism for the Kinnaridae in the islands (Table 4-8) and the number of species known from Puerto Rico (16) are remarkable.

Cicadelloidea

This intricate and difficult superfamily consists of about 17 families of the Auchenorrhyncha. Nonetheless, it can be argued that present knowledge of the taxonomy of the group is adequate for analysis. Young's (1977) monumental work on the taxonomy of the New World Cicadellinae is a fine example. In a similar way in which the Kinnaridae were selected to represent the Fulgoroidea, the Cicadellinae are chosen to represent the Cicadellidae.

Table 4-7. Distribution of Greater Antillean genera of Kinnaridae

Genus	North America	Mexico, Central America	South America	Cuba	Hispaniola	Jamaica	Puerto Rico
Atopocixius					+		
Dineparmene			+				
Lomagenes					+		
Microissus					+		
Occlidius	+	+		+		+	
Oreopenes							+
Paroeclidius			+				
Paraprosotropis							+
*Quilessa**							+

*Also known from the Lesser Antilles.

Table 4-8. Species level endemism of Greater Antillean Kinnaridae

Island	No. of species	No. of endemics	Percentage of endemism
Cuba	8	6	75
Hispaniola	8	8	100
Jamaica	3	1	33
Puerto Rico	16	16	100
Greater Antilles	35	31	89

The cicadelline fauna of the Greater Antilles consists of 42 species grouped into 11 genera (Table 4-9). Of this number of species, 5 are widespread and 37, or 88% are endemic. At the generic level endemism is 64%. Of the 11 recorded genera, only 4 are widespread in North, Central, and South America. Table 4-10 gives the percentage of endemism at the species level for the Cicadellini from each island.

Conclusion

A zoogeographic analysis of the Greater Antilles based on the Homoptera-Auchenorrhyncha reveals an extreme degree of endemism for each island individually, as well as for all four islands jointly. Although interisland faunistic relationships and affinities exist, maintaining them as a zoogeographic unit, each island's fauna has some individual peculiarities and identities. There is also evidence indicating that the Auchenorrhyncha of the Greater Antilles may be primarily derived from Central America and Mexico with little or no affinity to North or South America.

Table 4-9. Distribution of Greater Antillean genera of Cicadellinae

Genus	North America	Mexico, Central America	South America	Cuba	Hispaniola	Jamaica	Puerto Rico
Carneocephala	+	+	+	+	+	+	+
Ciminius	+	+	+	+			
Tylozygus	+	+	+	+	+	+	+
Hortensia	+	+	+	+	+	+	+
Bubacua				+			
Cubrasa				+			
Caribovia				+	+	+	+
Apogonalia				+	+	+	+
Camaija						+	
Hadria				+	+		
Ehagua					+	+	

Table 4-10. Species level endemism of Greater Antillean Cicadellinae

Island	No. of species	No. of endemics	Percentage of endemism
Cuba	23	15	65
Hispaniola	15	9	60
Jamaica	10	7	70
Puerto Rico	8	3	37
Greater Antilles	42	34	81

Interestingly, similar observations probably influenced the late great homopterist Z. P. Metcalf to propose (1947) the creation of a new zoogeographic region, the Caribbean, to include the West Indies, Mexico, and Central America, distinguishing it from the Neotropical region. This proposal should be the subject of research by modern zoogeographers.

References

Brown, F.M. 1978. The origins of the West Indian butterfly fauna. *In* Zoogeography in the Caribbean, pp. 5–30. The 1975 Leidy Medal Symposium. Special Publication No. 13, Acad. Sci. Philadelphia.

Caldwell, J.S. 1952. Review of the auchenorrhynchous Homoptera of Puerto Rico. Part II: The Fulgoroidea except Kinnaridae. J. Agric. Univ. of Puerto Rico 34:i–x, 133–269.

Caldwell, J.S., and L.F. Martorell. 1952. Review of the auchenorrhynchous Homoptera of Puerto Rico, part I: Cicadellidae. J. Agric. Univ. of Puerto Rico 34:i–viii, 1–132.

Deitz, L.L. 1975. Classification of the higher categories of the New World treehoppers (Homoptera: Membracidae). N.C. Agric. Exp. Sta., Tech. Bull. 225:i–iv, 1–177.

Dlabola, J. 1977. [Notes on the insect fauna of Cuba.] Casopis Nardniho Muzea 146 (1/4):151–156 [in Czech].

Fennah, R.G. 1942. New or little-known West Indian Kinnaridae (Homoptera: Fulgoroidea). Proc. Entomol. Soc. Washington 44(5):99–110.

———. 1945. Tropiduchidae and Kinnaridae from the Greater Antilles (Homoptera: Fulgoroidea). Psyche 52:119–138.

———. 1947. Two exotic new Fulgoroidea from the New World. Proc. Entomol. Soc. Washington 60:91–94.

———. 1948. New Pintaliine Cixiidae, Kinnaridae and Tropiduchidae from the Lesser Antilles (Homoptera: Fulgoroidea). Ann. and Mag. Nat. Hist., ser. 12, i, 417–437.

Hamilton, K.G.A. 1977. Review of the world species of *Prosapia* Fennah (Rhynchota: Homoptera: Cercopidae). Can. Entomol. 109:621–630.

Metcalf, Z.P. 1945. General Catalogue of the Homoptera, part 7: Kinnaridae, pp. 241–250. North Carolina State University, Raleigh.

———. 1947. The center of origin theory. Elisha Mitchell Sci. Soc. J. 62:149–175.

Metcalf, Z.P., and S.C. Bruner. 1925a. Notes and descriptions of the Cercopidae of Cuba. Psyche 32:95–105.

———. 1925b. Membracidae of Cuba. Bull. Brooklyn Entomol. Soc. 20:203–214.

———. 1930. Cuban Fulgorina. 1. The families Tropiduchidae and Acanaloniidae. Psyche 37:395–424.

———. 1936. The Cicadellidae of Cuba. J. Agric. Univ. Puerto Rico 20:915–979.

———. 1944. The Cercopidae of Cuba. Elisha Mitchell Sci. Soc. J. 60(2):109–127.

————. 1948. Cuban Flatidae with new species from adjacent regions. Ann. Entomol. Soc. America 41(1):63–118.

————. 1949. The Gyponidae and Ledridae of Cuba. Florida Entomol. 32(8):89–104.

O'Brien, L.B., and S.W. Wilson. 1985. Planthopper systematics and external morphology. In L.R. Nault and J.G. Rodriguez, eds., Leafhoppers and Planthoppers, pp. 61–102. John Wiley and Sons, New York.

Ramos, J.A. 1957. Review of the auchenorrhynchous Homoptera of Puerto Rico. J. Agric. Univ. Puerto Rico 41(10):38–117.

————. 1979. Membracidae de la República Dominicana (Homoptera: Auchenorrhyncha). Est. Exp. Agr. Univ. Puerto Rico Bol. 260:1–71.

————. 1983. Sinopsis de las cigarras de la República Dominicana. Carib. J. Sci. 19(1–2):61–68.

Van Duzee, E.P. 1907. Notes on Jamaican Hemiptera: A report on a collection of Hemiptera made on the island of Jamaica in the spring of 1906. Buffalo Soc. Nat. Sci. Bull. 8(5):3–79.

Wolda, H. 1984. Diversity and seasonality of Panamanian cicadas. Mitt. Schweiz. Ent. Ges. 57:451.

Young, D.A. 1977. Taxonomic study of the Cicadellinae (Homoptera: Cicadellidae), part 2: New World Cicadellini and the genus Cicadella. N.C. Agric. Exp. Sta. Tech. Bull. 239:i–vi, 1–1135.

5 · Kaleidoscopic Biogeography of West Indian Scaritinae (Coleoptera: Carabidae)

Stephen W. Nichols

The Scaritinae is a cosmopolitan group of ground beetles (Coleoptera: Carabidae) and includes about 1500 described species (Kryzhanovsky 1976). Considered one of the "primitive" groups of Carabidae (Erwin 1978) because its members possess a large suite of plesiomorphic character states (*sensu* Hennig 1966), the Scaritinae are recognized principally by their fossorially modified foretibiae and pedunculate union of prothorax and pterothorax.

Adult scaritines are burrowers, as implied by their morphology. Still, they are ecologically diverse; within the West Indian fauna alone there are intertidal halobionts, three types of halophobic hygrobionts, halophobic hygrophiles, and xerophiles (defined in Ecological Overview). In addition, there are differences in substrate preference; some species prefer gravel, others sand, and still others silt. This great ecological diversity is a major factor contributing to the mosaic biogeography of West Indian Scaritinae.

In this paper I summarize current knowledge about the biogeography of West Indian scaritines. A taxonomic revision of the West Indian Scaritinae, providing the data base for statements made here, will be published separately.

Study Area

For this study I defined the West Indian biogeographic region to include south Florida, the Yucatán Peninsula (in part), the Bahama Islands (including Turks and Caicos), the Greater Antilles, the Leeward Islands, the Windward Islands, and Barbados (Fig. 5-1). South Florida (Fig. 5-2) was defined to include that portion of peninsular Florida south of Lake Okeechobee, includ-

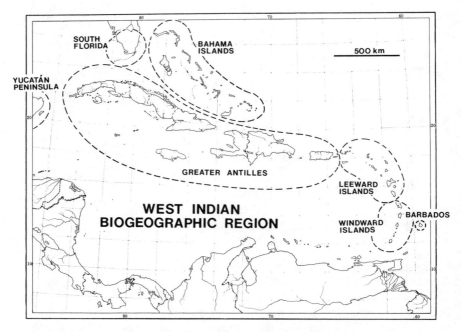

Figure 5-1. Map of the West Indian biogeographic region (see section "Study Area" for description).

ing seven counties: Lee, Hendry, Palm Beach, Collier, Broward, Monroe, and Dade. This area corresponds roughly to the subtropical area of Florida in the Holdridge system of life zones (Myers 1986:106, fig. 1). Part of the Yucatán Peninsula (Fig. 5-2)—the Mexican states of Yucatán and Quintana Roo (including Cozumel)—was also included; however, even larger portions of Middle America might also logically be included.

Taxonomic Overview

Sixty-five species of Scaritinae are recorded from the West Indian biogeographic region as defined above, although several species are probably introductions by man, and a few may represent waifs or even labeling errors. I regard the species inventory of the West Indian scaritine fauna to be fairly complete, with the probable exception of uncollected species with flightless adults (especially *Ardistomis*), inhabiting wet forests in relatively inaccessible areas in the mountains of Hispaniola, Cuba, and Jamaica.

Appendix 5.1 provides a distributional checklist of West Indian Scaritinae. Of the six tribes of Scaritinae present in the Western Hemisphere, all but one

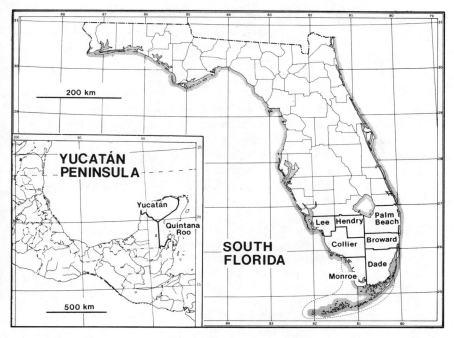

Figure 5-2. South Florida and the Yucatán Peninsula (in part) (see section "Study Area" for description).

(Salcediini) are represented in the West Indian biogeographic region (see Reichardt 1977). The distributional summary given for each species is a necessary caricature of the known distribution that is more accurately indicated by the data on specimen labels. Recently, Erwin and Sims (1984) published a checklist of the West Indian Scaritinae. My list differs considerably from theirs due to a broader definition of the West Indian biogeographic region, new records, several undescribed species, and much synonymy.

Ecological Overview

Most published information on the ecology of West Indian Scaritinae is contained in papers on West Indian Carabidae by Darlington (1934, 1935, 1936, 1937, 1939, 1941, 1947, 1953a, 1953b, 1970b) and in extralimital studies, for example, Blatchley (1910), Lindroth (1961), and Erwin (1981). Many specimens have been collected during short visits to the Caribbean by entomologists interested in general collecting; consequently, the infrequent ecological notes associated with pinned specimens are usually anecdotal.

Adult scaritines capable of flight are frequently taken at lights or in light traps, giving little idea of species' habitat requirements. In addition, adult carabids are rather long-lived (one to several years; Thiele 1977) so that their presence does not necessarily imply a breeding site, especially in the case of highly vagile dispersalists. For example, Erwin (1979) listed carabid species (none of these were scaritines) collected at lights on oil platforms in the Gulf of Mexico up to 160 km south of Morgan City, Louisiana.

In spite of the limitations mentioned above, two broad and biogeographically meaningful levels of descriptive categorization are possible for most species of West Indian Scaritinae, namely a species' station and a species' habitation. Both of these concepts were applied by A. P. de Candolle (1820) in discussing plant distributions (see Nelson 1978). With respect to stations and habitations Candolle writes:

> By the term *station* I mean the special nature of the locality in which each species customarily grows; and by the term *habitation* a general indication of the country wherein the plant is native. The term station refers essentially to climate, to the terrain of a given place; the term habitation relates to geographical, and even geological circumstances. The station of *Salicornia* is in salt marshes; that of the aquatic *Ranunculus* in stagnant fresh water. The habitation of both these plants is in Europe; that of the tulip tree, in North America. [Translation from Candolle 1820:383, in Nelson 1978:280]

Carabid beetles, with a few notable exceptions (including some specialized snail and seed feeders), are general predators, omnivores, or scavengers; consequently, their distributions are not generally limited by the availability of specific hosts. According to Thiele (1977:311), "Experiments and observations in the field have shown neither interspecific competition nor in fact any other biotic factors play a primary role in establishing the ecological pattern of distribution of carabids, this rather depending upon adaptation to abiotic factors." More recently, den Boer (1985) concluded that neither interspecific nor intraspecific competition are necessary conditions for evolutionary processes within carabid beetles, noting the lack of any convincing examples of biogeographic exclusion, character displacement, or niche separation attributable to competition.

Darlington (1943) categorized carabid beetles into three major ecological groups:

1. *geophiles*: ground-living forms inhabiting arid deserts, open plains, dry woods, or wet woods, without being associated with open water.
2. *hydrophiles*: ground-living Carabidae specifically associated with the banks of rapid brooks, slow rivers, or ponds, or with swamps, or just with very wet places, although none is completely aquatic.
3. *arboreal* species: species living above the ground.

Later, in his studies of the New Guinea carabid fauna, Darlington (1971: 173) introduced the term *mesophile*; however, defined as a species "found on the ground not associated with surface water" (Darlington 1971:154), *mesophile* had the same basic meaning as his term *geophile*. Darlington (1971: 173) also proposed a *fossorial* category (consisting of the Scaritinae); however, he indicated that "some fossorial species might be counted as hydrophiles or (a few) as mesophiles." Subsequent workers (e.g., Erwin 1979, 1981) have used mesophile to constitute a subset of Darlington's geophile concept, referring to relatively unspecialized forms, and have added additional specialized categories such as xerophile, ant symbiont, and so on.

Given the great ecological diversity exhibited by West Indian scaritines and the imprecision and confusion generated by the Darlington system, I have developed my own system, which relies largely on ecological principles and terminology adapted from Thiele's (1977) summary of carabid ecology. In Table 5-1 the West Indian Scaritinae are classified into four stations, defined relative to habitat preferences that focus on two abiotic parameters—relative humidity and the amount of sodium chloride in the environment. Thiele judged both of these factors to be very important in the ecological distribution of carabid beetles. I have abandoned the term *hydrophile* since it better describes amphibious organisms like frogs, salamanders, beavers, muskrats, hippopotamuses, and some breeds of domestic dog. Three substations of halophobic hygrobionts are also distinguished, representing habitats judged to differ in their long-term stability and differing in the beetles' exposure to sunlight. Riparian species inhabit the relatively least stable habitat with the greatest exposure to sunlight; humicoles inhabit the relatively most stable habitat with the least exposure to sunlight.

According to Nelson (1978:281), "Ecological biogeography is the study of stations; historical biogeography, the study of habitations." This is an oversimplification since it implies that historical biogeography is nonecological. Habitations (biogeographic areas in the sense of Nelson and Platnick 1981)

Table 5-1. Stations of West Indian Scaritinae

---

halobionts: hygrobiontic and hygrophilic species inhabiting areas with soil containing sodium chloride near the concentration of sea water; *intertidal* species in the West Indies.

halophobic hygrobionts: salt intolerant species inhabiting only areas with high relative humidity (close to 100%).
   riparian species: shoreline species inhabiting the wet banks of lotic or lentic waters.
   swamp-inhabiting species: species inhabiting freshwater marshes and swamps.
   humicoles: species found away from standing water in moist forest litter or humus.

halophobic hygrophiles: salt intolerant species that show a peak abundance in areas with high relative humidity; *eurytopic* species inhabiting a broad range of habitats including cultivated land.

xerophiles: species that show a peak abundance in areas with low relative humidity.

---

are to some degree ecologically delimited; if not, there would not be any barriers to dispersal. Even Europe and North America are ecologically defined by excluding marine habitats beyond the continental shelves.

The Holdridge system of life zones is a useful aid in describing the habitations of carabid beetles. This system utilizes four macroclimatic parameters: mean annual biotemperature (in °C), potential evapotranspiration ratio, average total annual precipitation (in mm), and a critical temperature line. Ewel and Whitmore (1973) have summarized some of the desirable features of this system. Erwin (1982) has applied this system to his studies of Central American carabids.

Habitations may be specified by combining Holdridge life zones with geographic regions, for example, lower montane wet—Luquillo Mountains, Puerto Rico, an area described and mapped by Ewel and Whitmore (1973). This system of naming habitations is particularly useful in regions with extreme topographic relief and distinct elevational zonation in climate. By defining habitations in this manner, one can apply the methods of vicariance biogeography (Nelson and Platnick 1981) and at the same time address one of Endler's main criticisms of vicariance biogeography—that it does not take into account ecological differences among taxonomic groups: "Vicariance biogeography ignores ecology, downgrades the importance of dispersal, and concentrates on the effects of speciation on the distribution of organisms. It is therefore exclusively concerned with the effects of history, and in particular the effects of geological changes which cause species to split, allowing the character distributions of the split populations to diverge" (Endler 1982: 350). Relevant areas that have been mapped in the Holdridge system include: peninsular Florida (Myers 1986), Haiti (OAS 1972), the Dominican Republic (Tasaico 1967), Jamaica (Gray and Symes 1969), Puerto Rico (Ewel and Whitmore 1973), and Venezuela (Ewel et al. 1968).

Table 5-2 presents a list of Holdridge life zones represented in the West Indies. The general temperature regime throughout the West Indies at lower elevations is subtropical, due in part to the northeast trade winds. Small, low islands like the Florida Keys are generally subtropical dry. Larger islands with higher elevations catch increasingly greater amounts of moisture and possess a broader array of life zones. West Indian islands are generally driest on the leeward side, that is, along the western or southern coasts. Montane wet forest, the coolest life zone in the West Indies, occurs only at high elevations in Cuba and Hispaniola; no Scaritinae are known from this life zone in the West Indies. The nearest areas on the mainland with similar climatic regimes include the Chiapan highlands of southern Mexico to the west and the southern Appalachian Mountains to the north; however, the southern Appalachians are very different in factors such as photoperiod and seasonal temperature variation. Granger (1985) has provided a general summary of Caribbean climatology.

Table 5-2. Holdridge life zones represented in the West Indies and approximately equivalent vegetation zones in the system of Beard (1949)

Holdridge life zone	Vegetation zone of Beard (1949)
Subtropical thorn woodland	——
Subtropical dry forest	Dry scrub—woodlands (in part)/deciduous seasonal forest
Subtropical moist forest	Dry scrub—woodlands (in part)/semi-evergreen and evergreen seasonal forest
Subtropical wet forest	Mostly "secondary forest"
Subtropical rain forest	Rain forest
Lower montane moist forest	——
Lower montane wet forest	Lower montane rain forest, palm-brake, and fern-brake
Lower montane rain forest	Montane thicket and elfin woodland
Montane wet forest	——

In the right-hand column of Table 5-2, I have attempted to homologize the vegetation zones of Beard (1949) that were developed to describe the natural vegetation of the Windward and Leeward Islands. This was done primarily by comparing Ewel and Whitmore's (1973) descriptions of Holdridge life zones of eastern Puerto Rico with Beard's (1949) descriptions of vegetation zones. Several Holdridge life zones that are apparently absent in the Lesser Antilles are represented in the Greater Antilles.

Biogeography by Tribe

Tribe PASIMACHINI (*New Status*)

Pasimachus (Figs. 5-3 and 5-4) are large, flightless xerophiles. In moist climates they tend to occupy well-drained sandy soil. There are about 25 species, with the genus being restricted to North America (Bänninger 1950). The nearest relative of *Pasimachus* is probably the genus *Mouhotia*, which is distributed in eastern Asia (Japan and the Indo-Malayan region). At least three species of *Pasimachus* (subg. *Pasimachus*) range into south Florida, all three having been found on Key West. On the Yucatán Peninsula the genus is represented by a single species of the subgenus *Emydopterus*, *P. purpuratus*. No *Pasimachus* have been recorded from the West Indian islands, although suitable habitat appears to be widespread.

Given the ecology, flightlessness, and relatively large body size of *Pasimachus*, it is unlikely that any species would be blown, rafted, or introduced across the Straits of Florida or the Straits of Yucatán of today. Barring extinction, the distribution of *Pasimachus* may be regarded as evidence against land connections between mainland North America and the Greater

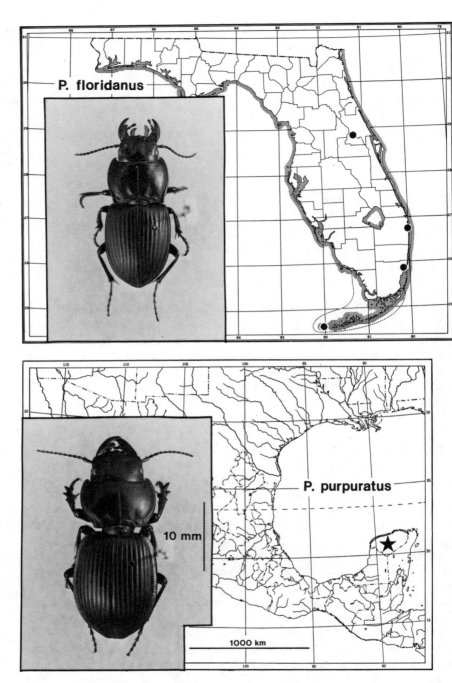

Figure 5-3. West Indian *Pasimachus* I: *P. floridanus* (specimen from Key West, Fla.; U.S. National Museum, Washington, D.C.) and *P. purpuratus* (specimen from Mexico; Museum of Comparative Zoology, Harvard Univ., Cambridge, Mass.). Both specimens photographed at the same scale.

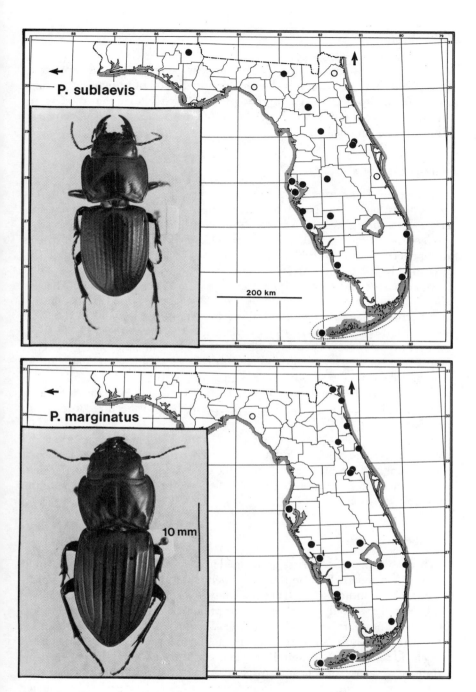

Figure 5-4. West Indian *Pasimachus* II: *P. sublaevis* (specimen from Lake Worth, Fla.; Casey Collection, U.S. National Museum, Washington, D.C.), and *P. marginatus* (specimen from Punta Gorda, Fla.; U.S. National Museum, Washington, D.C.). Open circles refer to county records. Both specimens photographed at the same scale.

Antilles (at least since Cuba became permanently emergent), thus supporting Brigg's (1984) conclusion based upon the distribution of primary freshwater fishes. Darlington (1938, 1970b) had previously concluded that the Greater Antillean carabid fauna was derived through overwater dispersal, primarily from Middle America.

Tribe DYSCHIRIINI

The genus *Dyschirius* is cosmopolitan, mostly inhabiting temperate climates, with numerous North American species (Lindroth 1961). Four species of *Dyschirius* are known from the West Indian biogeographic region. Two of these (Fig. 5-5), *D. pumilus* and *D. abbreviatus*, are riparian species that penetrate into south Florida but are not known from the Florida Keys, presumably due to the scarcity of standing fresh water.

D. sublaevis and *D.* nr. *erythrocerus* (Fig. 5-6) are halobionts. There is some question whether *D. erythrocerus* LeConte (described from Ohio) is conspecific with the form found along the Atlantic coast and in the West Indies (Lindroth 1961:147–148). Whitehead (1969b) reported *D. sublaevis* and *D.* nr. *erythrocerus* from sea beaches and salt marshes along the Gulf coast. Both species are recorded here from the Florida Keys, Cuba, and the Yucatán Peninsula, and both species may prove to be circum-Gulf. *D.* nr. *erythrocerus* is also known from the Bahama Islands and Hispaniola. In spite of much collecting on Puerto Rico, the species has not been recorded from that island.

The genus *Dyschirius* suggests the importance of a species' station to its ability to colonize the West Indian islands. There is a much greater chance of a halobiont finding suitable habitat on the other side of an ocean barrier than a riparian species. A halobiont would also be more suited to surviving the journey by rafting.

Tribe SCARITINI

Three subgenera of *Scarites* (as defined in Reichardt 1977) are represented in the West Indies. The nominate subgenus is predominantly Pantropical with five West Indian species (Figs. 5-7 and 5-8). The large *S. quadriceps* (Fig. 5-7) is a fully winged, halophobic hygrophile that ranges into south Florida but is not known from the Florida Keys. *S. subterraneus* (Fig. 5-8), another fully winged halophobic hygrophile, ranges farther south, being known from the Florida Keys, the Yucatán Peninsula, and throughout Cuba in subtropical moist and subtropical dry lowlands. Very similar in form but distinctly larger than *S. subterraneus*, *S. alternans* appears to be restricted to higher ground in the Trinidad Mountains of Las Villas Province and in Oriente Province of Cuba; although fully winged, this species may have a disjunct

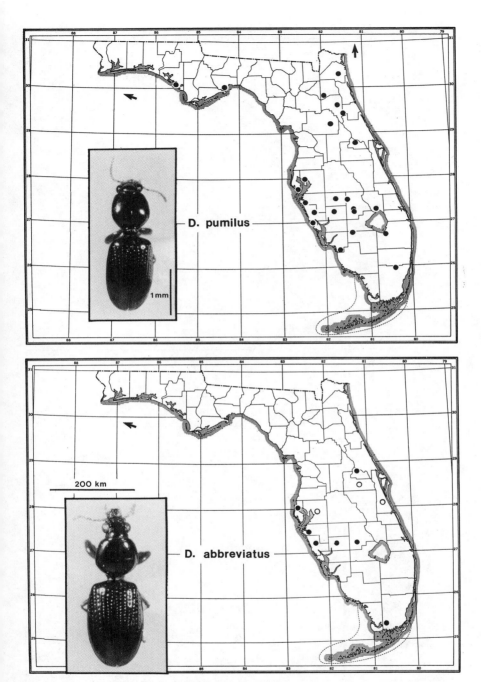

Figure 5-5. West Indian *Dyschirius* I: *D. pumilus* (specimen from Dunedin, Fla.; H. C. Fall Collection, Museum of Comparative Zoology, Harvard Univ., Cambridge, Mass.), and *D. abbreviatus* (specimen from Everglades National Park, Fla.; Florida Dept. of Agriculture, Gainesville). Open circles in lower map refer to county records. Both specimens photographed at the same scale.

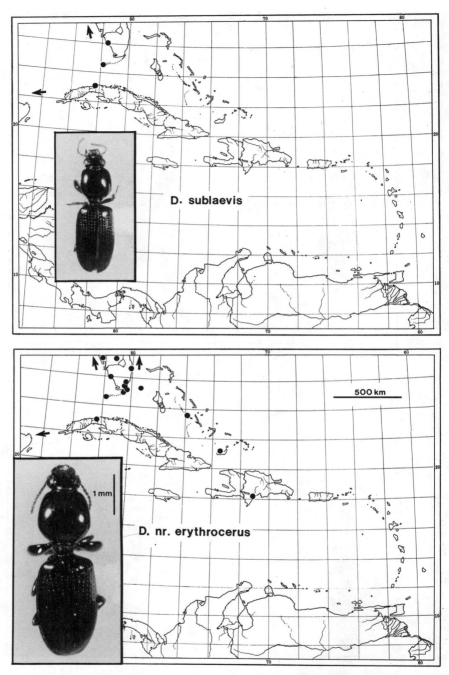

Figure 5-6. West Indian *Dyschirius* II: *D. sublaevis* (specimen from Dunedin, Fla.; H. C. Fall Collection, Museum of Comparative Zoology, Harvard Univ., Cambridge, Mass.), and *D.* nr. *erythrocerus* (specimen from South Bimini Island, Bahamas; American Museum of Natural History, New York). Both specimens photographed at the same scale.

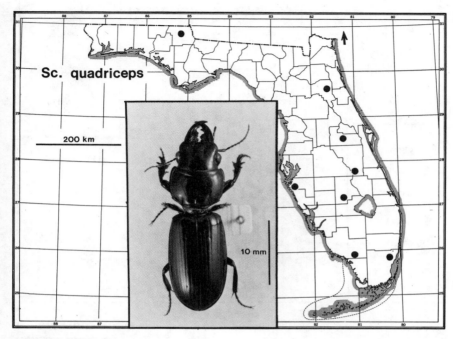

Figure 5-7. West Indian *Scarites*: subgenus *Scarites* I: *S. quadriceps* (specimen from Orlando, Fla.; Cornell Univ. Insect Collections, Ithaca, N.Y.).

range. The presence of *S. subterraneus* on Cuba could represent a relatively recent invasion or even an introduction, since the species is commonly found in agricultural settings. The occurrence of *S. alternans* and *S. subterraneus* together on Cuba may represent an example of double invasion (Williamson 1981:156–157). MacLean and Holt (1979) discuss a potentially similar relationship between two species of *Sphaerodactylus* lizard on the island of St. Croix, with the invading species occupying most of the island and the resident species restricted to several disjunct areas.

    S. illustris is a tropical species whose range extends north as far as Tlacotalpan in the state of Veracruz, Mexico. The species is known from the Swan Islands off the coast of Honduras and is recorded from Cuba but with no specific locality. *S. illustris* ranges south through Central America and into South America. *S. marinus* is a brachypterous, intertidal halobiont commonly found among seaweed washed up on sandy beaches (Nichols 1986a). Rafting of this species is likely and is favored by south-to-north surface currents through the straits of Yucatán and west-to-east surface currents through the Straits of Florida. A single specimen has been collected from Louisiana.

    One species belonging to the Pantropical subgenus *Taeniolobus*, *S.*

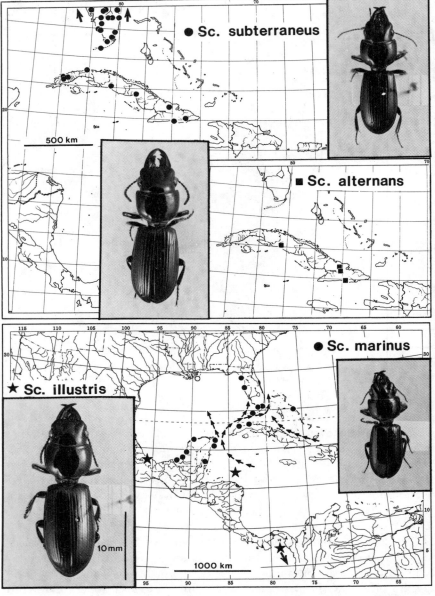

Figure 5-8. West Indian *Scarites*: subgenus *Scarites* II: *S. subterraneus* (specimen from Lucedale, Miss.; Cornell Univ. Insect Collections, Ithaca, N.Y.); *S. alternans* (specimen from Holquín, Cuba; Museum of Comparative Zoology, Harvard Univ., Cambridge, Mass.); *S. illustris* (specimen from Swan Islands, Honduras; Museum of Comparative Zoology, Harvard Univ.); and *S. marinus* (specimen from Cape Sable, Fla.; Florida Dept. of Agriculture, Gainesville). All specimens photographed at the same scale. Series of arrows indicate the general direction of surface currents around Cuba. Open star in lower map stands for nonspecific locality record of *S. illustris* from Cuba. Open circle in lower map stands for nonspecific locality record of *S. marinus* from the state of Louisiana.

cubanus, inhabits the Trinidad Mountains of central Cuba (Bänninger 1937) (Fig. 5-7). *S. cubanus* is a halophobic hygrobiont, and the habitation of this species is lower montane wet—Trinidad Mountains, Cuba. The nearest described relative of *S. cubanus* is the Colombian *S. lebasi*, based upon the shared derived form of the labial mentum, which is deeply depressed medially (Bänninger 1941); however, there is an undescribed species that inhabits southern Mexico (George E. Ball, University of Alberta, pers. comm.) that should be studied before any hypothesis is formed regarding the origin of *S. cubanus*. All described members of the subgenus *Taeniolobus* are flightless.

The Pantropical subgenus *Distichus* is represented by two species in the West Indies (Fig. 5-9). Both of these species possess anomalous distributions. The Middle American *S. octocoelus* (not equal to *S. octocoelus* in the sense of Bänninger 1938) is also known from Jamaica and Guadeloupe. The South American *S. orientalis* is recorded from Port-au-Prince, Haiti; however, this species was not encountered by Philip Darlington when he collected in Haiti (Darlington 1936). For both species the island occurrences probably represent introductions (or possibly mislabeling) by man. Specimens of both species from the islands are fully winged. While *S. octocoelus* is a relatively large insect, it was not recorded from Guadeloupe by Fleutiaux and Sallé (1889). Both species are known from moist and dry regions; however, *S. octocoelus* is largely subtropical and *S. orientalis* is largely tropical. Consequently, I would predict that even if *S. orientalis* has occasionally been introduced to Haiti, it probably has not become established there. One might argue a case for deleting *S. orientalis* from the list of West Indian scaritines or for treating the species as a waif; however, to make the present biogeographic analysis valid I have treated all records based upon my own examination of specimens.

Antilliscaris (new status) (Fig. 5-10), revised by Hlavac (1969), is one of two genera of the family Carabidae endemic to the West Indies. The other endemic genus is *Barylaus* (tribe Pterostichini: Caelostomina) (Liebherr 1986). *Antilliscaris* includes four nominal species: the giant *A. megacephalus* from the Luquillo Mountains of eastern Puerto Rico; the considerably smaller *A. mutchleri*, also from the Luquillo Mountains; *A. danforthi* from the Maricao Forest of western Puerto Rico; and *A. darlingtoni* from Mount Basil in Haiti, the latter described and still known only from fragments of two individuals. The conspicuous size difference between the two sympatric species, *A. megacephalus* and *A. mutchleri*, is about what one would predict if *A. megacephalus* had gained one larval instar, assuming one started with the typical carabid number of three and applied growth indexes published for other scaritine taxa by van Emden (1942).

Antilliscaris are flightless and inhabit lower montane wet and lower montane rain forest (see Ewel and Whitmore 1973 and OAS 1972). The four species constitute a monophyletic group based upon the shared loss of the

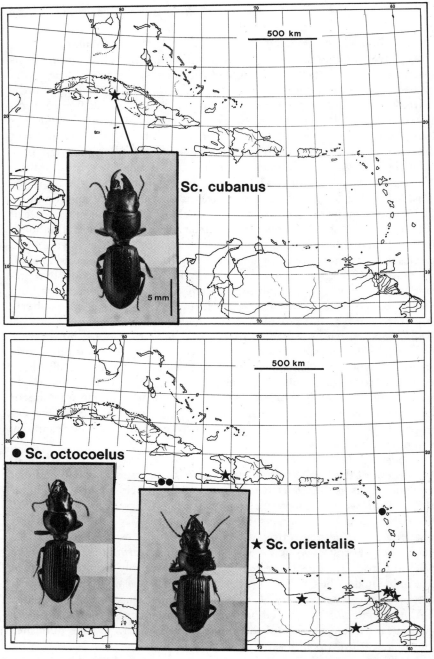

Figure 5-9. West Indian *Scarites*: subgenus *Taeniolobus* (upper map): *S. cubanus* (specimen from Trinidad Mountains, Cuba; Museum of Comparative Zoology, Harvard Univ., Cambridge, Mass.); subgenus *Distichus* (lower map): *S. octocoelus* (specimen from Sainte Anne, Guadeloupe; Muséum National d'Histoire Naturelle, Paris), and *S. orientalis* (specimen from Port-of-Spain, Trinidad; S. W. Nichols Collection, Ithaca, N.Y.). All specimens photographed at the same scale.

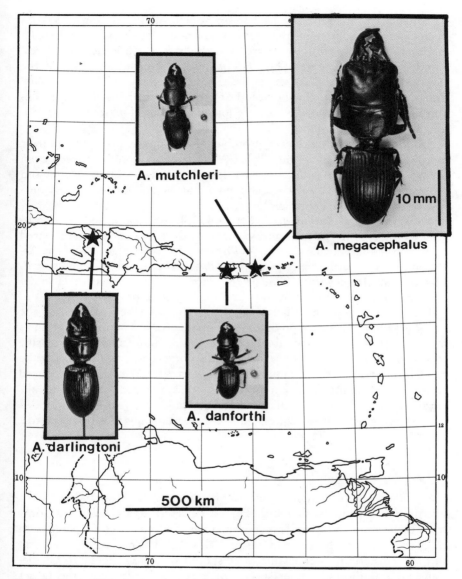

Figure 5-10. *Antilliscaris*: *A. megacephalus* (specimen from Mount El Yunque, Puerto Rico; Strickland Museum, Univ. of Alberta, Edmonton); *A. mutchleri* (specimen from Mount El Yunque, Puerto Rico; Museum of Comparative Zoology, Harvard Univ., Cambridge, Mass.); *A. danforthi* (specimen from the Maricao Forest, Puerto Rico; J. A. Ramos Collection, Mayagüez, P.R.); and *A. darlingtoni* (specimen from Mount Basil, Haiti; Museum of Comparative Zoology, Harvard Univ.). All specimens photographed at the same scale.

elytral-epimeral wing-locking mechanism (see Hlavac 1969), which, as far as I can determine, is present in all other scaritines. Loss of this wing-locking mechanism is not directly correlated with flight loss, since there are "old" flightless scaritine lineages (e.g., *Pasimachus* of North America and the subtribe Storthodontina of Madagascar; see Basilewsky 1973 for a discussion of the latter group) that possess the elytral-epimeral wing-locking mechanism. The elytra of *Antilliscaris* are joined together along the suture in a tongue-and-groove fashion.

Antilliscaris are halophobic hygrobionts. I would guess, based upon their morphology, that they are humicoles. Although undoubtedly derived from fossorial ancestors, *Antilliscaris* have lost the strongly dilated profemora possessed by their near relatives, implying that they no longer actively burrow or that they move through loose substrate (Hlavac 1969).

A study of larval *Antilliscaris* (Nichols 1986b) was conducted in order to gather insights into the relationships of *Antilliscaris* to other genera. Based upon selected larval characters, *Antilliscaris* groups with genera of the Afrotropical Region, most specifically with *Prodyscherus*, which is endemic to Madagascar (Basilewsky 1973, 1976). However, many genera lack described larvae, thus tempering any far-flung biogeographic hypotheses.

The abundance of these beetles may have been severely affected by the introduction of nonnative predators and omnivores to Puerto Rico, such as feral cats (*Felis catus*), mongooses (*Herpestes auropunctatus*), the marine toad (*Bufo marinus*), the wharf rat (*Rattus norvegicus*), and especially the black roof rat (*R. rattus*). Diamond (1985) pointed out the role of introduced rats as causal agents in the extermination of island birds, small mammals, large insects, molluscs (Hawaiian achatinellid land snails) and cold-blooded vertebrates (most ground snakes and lizards of Mauritius plus frogs, lizards, and tuataras of New Zealand).

Tribe FORCIPATORINI

Stratiotes is a Neotropical genus represented in the West Indies by the Martinique endemic, *S. iracundus* (Fig. 5-11), described and still known only from the unique holotype. This specimen has reduced eyes (compared with other species of *Stratiotes*), and it is probably flightless based upon the form of the metasternum. Besides the three species that are endemic to Guadeloupe, this is the only other species of Scaritinae endemic to the Lesser Antilles. A sister group to this species has not been identified, and nothing is known about the species' ecology.

Like *Stratiotes*, *Camptodontus* is a Neotropical genus. *C. anglicanus* (Fig. 5-11), described from a specimen found dead in a market area east of Peckham, England (Stephens 1827), is known from Dominica, having possibly been accidentally introduced to Dominica from South America. This

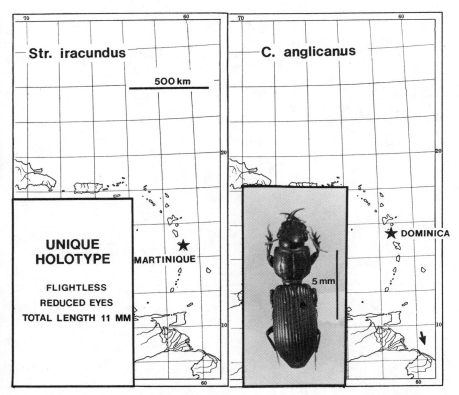

Figure 5-11. West Indian *Stratiotes* and *Camptodontus*: *S. iracundus* (left map) (unique holotype; Chaudoir-Oberthur Collection, Muséum National d'Histoire Naturelle, Paris), and *C. anglicanus* (right map) (specimen from Springfield Estate, Dominica; Museum of Comparative Zoology, Harvard Univ., Cambridge, Mass.).

fully winged species inhabits subtropical wet forest on Dominica, and collection records indicate that the species is established on the island.

Tribe CLIVININI

About two-thirds of the West Indian species of Scaritinae belong to the tribe Clivinini. *Halocoryza arenaria* (Fig. 5-12) is a minute, intertidal halobiont. Widespread throughout the Caribbean, this species is also recorded from the Gulf of Biafra of Africa (Bruneau de Miré 1979) and from the state of Pernambuco, Brazil (Nichols, unpubl. data). Members of the genus *Halocoryza* are Neotropical and Afrotropical, and a second New World species, *H. acapulcana*, inhabits the Pacific side of Middle America (Whitehead 1966, 1969a, 1979, Basilewsky 1973). *H. arenaria* is polymorphic in hind-wing condition, varying from brachypterous to fully winged (Whitehead 1969a).

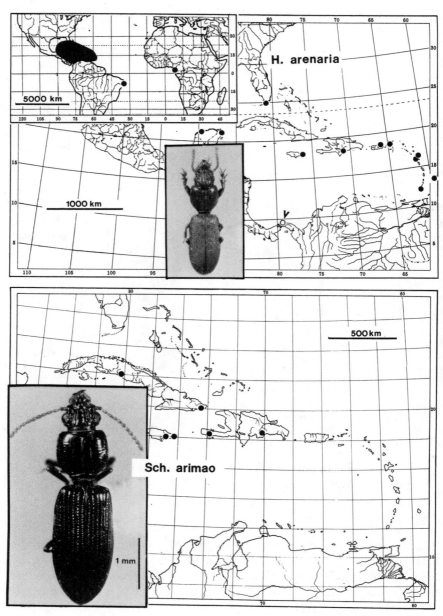

Figure 5-12. West Indian *Halocoryza* and *Schizogenius*: *H. arenaria* (upper map) (specimen from Pernambuco, Brazil; British Museum of Natural History, London), and *S. arimao* (lower map) (specimen from Soledad nr. Cienfuegos, Cuba; Museum of Comparative Zoology, Harvard Univ., Cambridge, Mass.). Both specimens photographed at the same scale. Open circle in upper map denotes literature record of *H. arenaria* from the northern coast of Panama (Bruneau de Miré 1979:329).

Schizogenius arimao (Fig. 5-12) is endemic to the Greater Antilles (Cuba, Jamaica, and Hispaniola). This is a riparian species typically inhabiting gravel shorelines of rivers and streams (see Dunn 1986). Whitehead (1972) hypothesized that this species is the sister group of a two-species clade; one of these species occurs in Central America, the other in South America. *Schizogenius* are restricted to the Western Hemisphere.

Four subgenera of the genus *Clivina* are represented in the West Indies. *C. limbipennis* (Fig. 5-11) is endemic to the Greater Antilles (Cuba, Hispaniola, and Puerto Rico) and belongs to the subgenus *Isoclivina* which is cosmopolitan, excluding South America (Kult 1959). *C. planicollis*, which inhabits the southeastern United States (Whitehead 1974) and eastern Mexico (Bates 1881), is probably the nearest relative of *C. limbipennis*; however, a world-level study of *Isoclivina* is required before a credible cladogenetic hypothesis can be constructed.

The *Clivina* subgenus *Reichardtula* (Whitehead, in Reichardt 1977:391) (= *Eupalamus* of Kult 1959) is cosmopolitan, excluding South America and Australia. *C. americana* (Fig. 5-13) is a swamp-inhabiting species; *C. acuducta* is a riparian species. Both taxa penetrate south Florida from the north, the former also being known from the Bimini Islands.

Five riparian or swamp-inhabiting species of the Neotropical (marginally Nearctic) subgenus *Semiclivina* (see Kult 1947) are known from the West Indies (Fig. 5-14). Two of these, *C. cubae* (Cuba) and *C. addita* (Hispaniola and Puerto Rico), are endemic to the West Indies. The two species known from the Lesser Antilles, *C. elongata* and *C. oblita*, may represent waifs from South America (both species are also known from northern South America) since they are known from the West Indies by very few specimens.

Members of the subgenus *Paraclivina* are Neotropical (marginally Nearctic) and Australian (Kult 1947), with one species apparently having been introduced to the South Pacific (see below). Seven species belonging to the subgenus *Paraclivina* are known from the West Indies (Figs. 5-15 and 5-16). With the possible addition of one species (*C.* nr. *latiuscula*), only *C. biguttata* is endemic to the West Indies. Further study is needed to determine the taxonomic status of *C.* nr. *latiuscula* (*C. latiuscula* is a species described from the Amazon Basin of South America). *C. fasciata* (Fig. 5-15) is circum-Caribbean and is also known, probably via introduction, from the Philippine Islands (Luzon, Leyte, Mindoro, Negros) and the Mariana Islands (Guam, Saipan) (Darlington 1970a:11–12). Darlington (1970a:6, 36) reported another similar probable introduction of the Middle American carabid *Selenophorus pyritosus* (Carabidae: Harpalini) to Tuamotus, Tahiti, Raiatea, and Guam. In both cases Darlington suggested introduction via ballast from Spanish galleons. *C. tuberculata* and *C. marginipennis* (Fig. 5-16) are possible introductions from South America. *C. tuberculata* was described from Bogotá, Colombia, and I have also seen specimens from Catamarca and

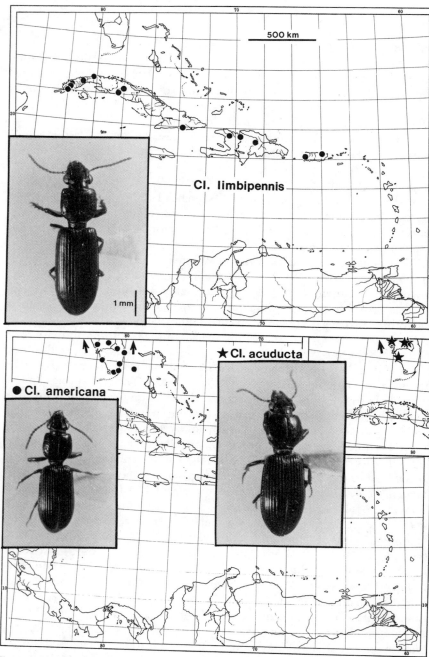

Figure 5-13. West Indian *Clivina*: subgenus *Isoclivina*: *C. limbipennis* (upper map) (specimen from San Vicente, Pinar del Río, Cuba; Museum of Comparative Zoology, Harvard Univ., Cambridge, Mass.); subgenus *Reichardtula* (lower map): *C. americana* (specimen from South Bimini Island, Bahamas; Museum of Comparative Zoology, Harvard Univ.), and *C. acuducta* (specimen from Cortland County, N.Y.; Cornell Univ. Insect Collections, Ithaca, N.Y.). All specimens photographed at the same scale.

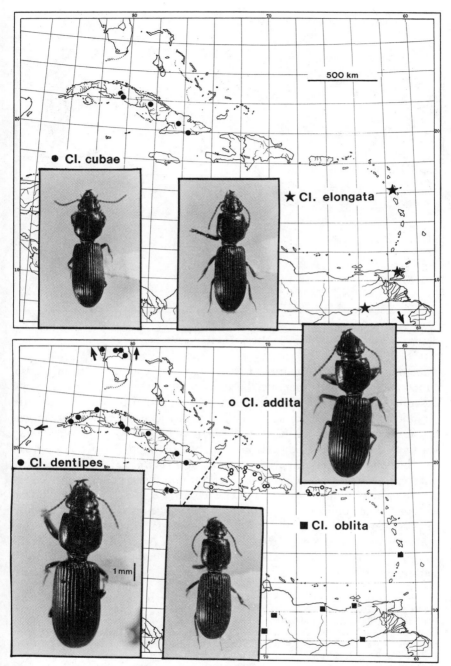

Figure 5-14. West Indian *Clivina*: subgenus *Semiclivina*: *C. cubae* (specimen from Cauto El Cristo, Oriente Province, Cuba; Museum of Comparative Zoology, Harvard Univ., Cambridge, Mass.); *C. elongata* (specimen from C. des Mamelles, Guadeloupe; Muséum National d'Histoire Naturelle, Paris); *C. addita* (specimen from Ennery, Haiti; Museum of Comparative Zoology, Harvard Univ.); *C. dentipes* (specimen from Cayamas, Cuba; U.S. National Museum, Washington, D.C.); and *C. oblita* (specimen from Hope LT., British Guiana; U.S. National Museum, Washington, D.C.). All specimens photographed at the same scale.

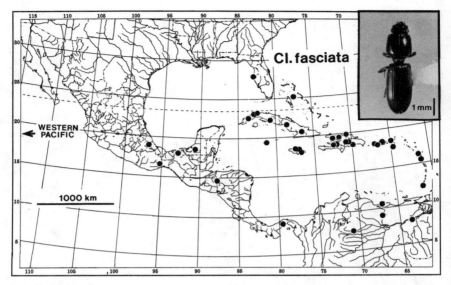

Figure 5-15. West Indian *Clivina*: subgenus *Paraclivina* I: *C. fasciata* (specimen from Los Hidalgos, Dominican Republic; U.S. National Museum, Washington, D.C.).

Tucumán provinces, Argentina. I have examined a specimen from Viçosa, Minas Gerais, Brazil, that fits the description of *C. fuscicornis* Putzeys, described from Campos, Rio de Janeiro, Brazil, and also recorded from Rio Grande do Sul, Brazil; this specimen is virtually identical to *C. marginipennis*. *C. marginipennis* and *C. fuscicornis* may be synonymous. Nickle and Castner (1984) discussed the introduction of three South American species of soil-dwelling mole crickets, *Scapteriscus abbreviatus*, *S. imitatus*, and *S. didactylus* (Orthoptera: Gryllotalpidae), into the West Indies. The mode of transport that brought these mole crickets to the West Indies would also explain the introduction of members of the subgenus *Paraclivina*. Nichols (1985) reported the probable introduction of *C. (Semiclivina) vespertina* to the southern United States from southern South America. Members of the subgenus *Paraclivina* are eurytopic hygrophiles and commonly associated with agriculture. *C. fasciata* is the most commonly encountered scaritine in

Figure 5-16. West Indian *Clivina*: subgenus *Paraclivina* II: *C. marginipennis* (specimen from Port d'Enfer, Portland, Guadeloupe; Muséum National d'Histoire Naturelle, Paris); *C. bipustulata* (specimen from Cayamas, Cuba; U.S. National Museum, Washington, D.C.); *C.* nr. *latiuscula* (specimen from Guadeloupe; Museum of Comparative Zoology, Harvard Univ., Cambridge, Mass.); *C. tuberculata* (specimen from Codrington College, Barbados; British Museum of Natural History, London); *C. biguttata* (specimen from Aguadores, Oriente Province, Cuba; Museum of Comparative Zoology, Harvard Univ.); and *C. tristis* (specimen from Dos Poos, Bonaire; Zöologisch Museum, Universiteit van Amsterdam). All specimens photographed at the same scale.

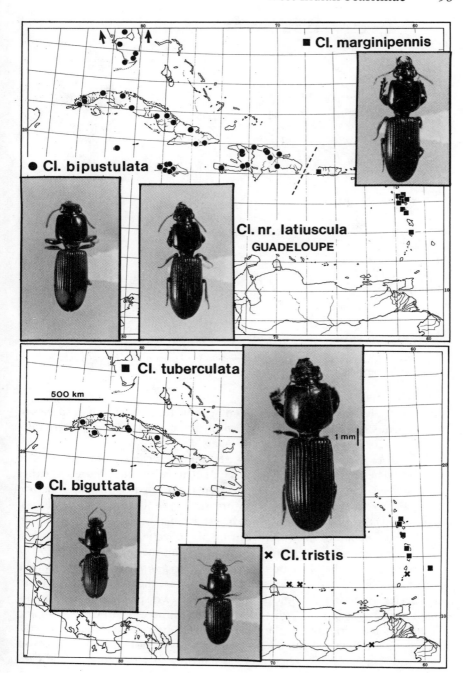

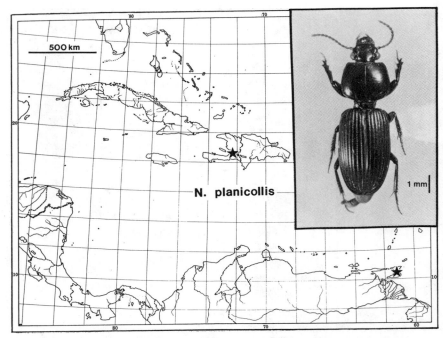

Figure 5-17. West Indian *Nyctosyles*: *N. planicollis* (specimen from Trinidad; U.S. National Museum, Washington, D.C.).

the West Indies, frequently being taken in light traps. West Indian species of the subgenus *Paraclivina* inhabit subtropical moist regions and all are fully winged.

Nyctosyles planicollis (Fig. 5-17) is a South American species. This species is recorded here from Port-au-Prince, Haiti. This record may represent either a mislabeled specimen or an introduction from northern South America. Being a tropical species like *Scarites orientalis* (see the above discussion of that species), *N. planicollis* may not be established on Hispaniola. Darlington did not encounter this species during his collecting in Haiti (Darlington 1936). *Nyctosyles* includes three South American species (Reichardt 1977).

Four species of the Neotropical genus *Oxydrepanus* are known from the West Indies (Fig. 5-18). *O. rufus* and *O. micans* are flighted, swamp-inhabiting species found in subtropical lowlands. The two species possess complementary distributions. *O. rufus* is distributed in the western Caribbean and *O. micans* is distributed in the eastern Caribbean. *O. reicheoides* and *O. coamensis* are flightless humicoles found in lower montane wet and lower montane rain forest. *O. reicheoides* is endemic to Hispaniola in the Cordillera Central; *O. coamensis* is endemic to Puerto Rico.

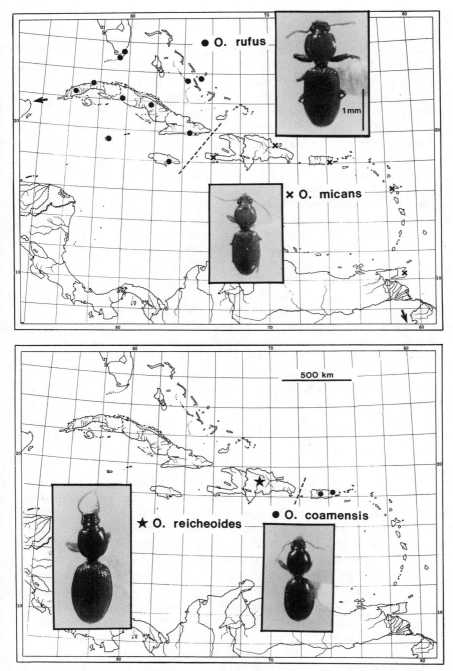

Figure 5-18. West Indian *Oxydrepanus*: *O. rufus* (specimen from Cayamas, Cuba; U.S. National Museum, Washington, D.C.); *O. micans* (specimen from Humacao, Puerto Rico; Museum of Comparative Zoology, Harvard Univ., Cambridge, Mass.); *O. reicheoides* (specimen from Loma Rucilla or mountains north, Dominican Republic; Museum of Comparative Zoology, Harvard Univ.); and *O. coamensis* (specimen from Mount El Yunque, Puerto Rico; Museum of Comparative Zoology, Harvard Univ.). All specimens photographed at the same scale.

Like the subgenus *Paraclivina* of *Clivina*, members of the Neotropical (marginally Nearctic) genus *Aspidoglossa* are fully winged, eurytopic hygrophiles commonly found in lowland subtropical moist areas throughout the West Indies (Fig. 5-19). None of the four species in the West Indies are endemic to the region. Like *Oxydrepanus rufus* and *O. micans*, the three larger species exhibit an interesting complementarity in their distributions, not clearly attributable to sea barriers or specific differences in ecology.

Four species of the Neotropical (marginally Nearctic) genus *Semiardistomis* (new status) inhabit the West Indies (Fig. 5-20). *S. viridis* and *S. cyaneolimbatus* are riparian, with the latter species being endemic to the region. *S. puncticollis* and *S. laevistriatus* are humicoles, with the latter species being endemic to Guadeloupe. *S. laevistriatus* is brachypterous; the other three species of *Semiardistomis* are fully winged.

The genus *Ardistomis* is notable for its high degree of endemicity, being the most speciose genus in the West Indies. This Neotropical (marginally Nearctic) genus is represented in the West Indies by three major lineages. The first of these (Fig. 5-21) includes six flightless species, all of which are extremely restricted in distribution. These are all humicoles found in lower montane wet and lower montane rain forest at higher elevations. The species include *A. ramsdeni* from Pico Turquino in the Sierra Maestra of Cuba; *A. franki* (manuscript name) from Hardwar Gap in the Blue Mountains of Jamaica; *A. alticolus* from Mount Bourette of the La Selle massif in Haiti; *A. nigroclarus* and *A. rufoclarus* from the Cordillera Central of the Dominican Republic; and *A. guadeloupensis* from La Soufrière on Basse Terre of Guadeloupe. Peck (1975) reported the occurrence of two species of blind, troglobitic *Ardistomis* from Falling Cave (0.4 km E Douglas Castle, 6.4 km NW Kellits; St. Ann Parish) and Pedro Great Cave 1.0 km SE Pedro River; St. Catherine Parish) on Jamaica. I have not been able to examine those specimens but suspect that they are true *Ardistomis*, probably belonging to this first group.

A. schaumi and *A. elongatulus* are members of the second lineage (Fig. 5-22, top map). *A. schaumi* is circum-Gulf, ranging from Florida to Honduras; *A. elongatulus* is endemic to Cuba. These two species and the remaining *Ardistomis* to be discussed are fully winged, marsh- and swamp-inhabiting species, found predominantly in lowland subtropical areas.

The third lineage (Fig. 5-22, bottom map) includes five species distributed across the Caribbean arc. There are west-to-east progressions in two characters: (1) reduction in the genital armature of the male internal sac, and (2) the loss of preapical elytral spots. *A. obliquatus* and *A. nitidipennis* possess preapical spots. *A. hispaniolensis* (manuscript name) is polymorphic for this trait, and there is an apparent cline from western to eastern Hispaniola; western individuals are predominantly spotted, eastern individuals predominantly unspotted. *A. mannerheimi* and *A. atripennis* lack preapical spots.

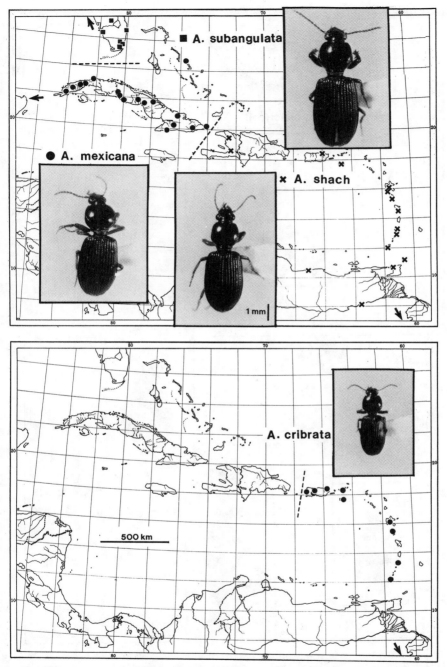

Figure 5-19. West Indian *Aspidoglossa*: *A. subangulata* (specimen from L. S. Walker State Park, Ware County, Ga.; Cornell Univ. Insect Collections, Ithaca, N.Y.); *A. mexicana* (specimen from Cayamas, Cuba; U.S. National Museum, Washington, D.C.); *A. schach* (specimen from Bacolet Point, Tobago; Museum of Comparative Zoology, Harvard Univ., Cambridge, Mass.); and *A. cribrata* (specimen from Grenville, Grenada; British Museum of Natural History, London). All specimens photographed at the same scale.

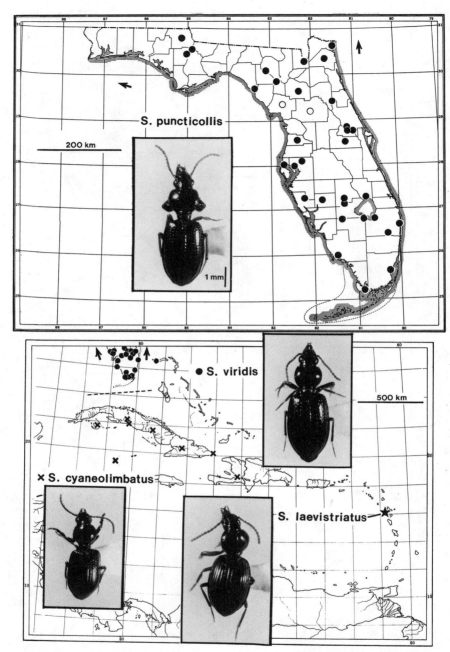

Figure 5-20. West Indian *Semiardistomis*: *S. puncticollis* (specimen from Torreya State Park, Liberty County, Fla.; Strickland Museum, Univ. of Alberta, Edmonton); *S. viridis* (specimen from Tuscaloosa, Ala.; Strickland Museum, Univ. of Alberta, Edmonton); *S. cyaneolimbatus* (specimen from Isla de Pinos, Cuba; Field Museum of Natural History, Chicago); and *S. laevistriatus* (specimen from Baines Jaunes, Guadeloupe; U.S. National Museum, Washington, D.C.). Open circles in upper map refer to county records. All specimens photographed at the same scale.

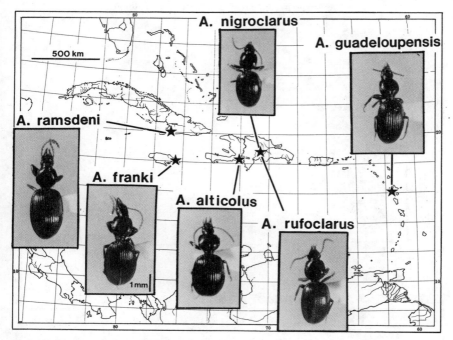

Figure 5-21. West Indian *Ardistomis* I: *A. nigroclarus* (specimen from the vicinity of Valle Nuevo nr. Constanza, Dominican Republic; Strickland Museum, Univ. of Alberta, Edmonton); *A. guadeloupensis* (specimen from Guadeloupe; Hungarian Natural History Museum, Budapest); *A. ramsdeni* (specimen from Pico Turquino, Cuba; Museum of Comparative Zoology, Harvard Univ., Cambridge, Mass.); *A. franki* (manuscript name) (specimen from Hardwar Gap, Jamaica; Strickland Museum, Univ. of Alberta, Edmonton); *A. alticolus* (specimen from Morne Bourette nr. Furey, Haiti; Museum of Comparative Zoology, Harvard Univ.); and *A. rufoclarus* (specimen from vicinity of Valle Nuevo nr. Constanza, Dominican Republic; Museum of Comparative Zoology, Harvard Univ.). All specimens photographed at the same scale.

Biogeographic Analysis

Recognizing the relatively small number of taxa (65 species) treated here, I have attempted to provide an alpha-biogeographic summary in Figure 5-23. The first number under each land area refers to the number of species recorded from that area. I included all species in this total regardless of whether some species might be regarded as waifs, provided that I had examined specimens labeled as originating from that area. The percentage that follows indicates the percentage of that area's fauna that is endemic. In calculating these percentages I treated *Ardistomis mannerheimi* as a Puerto Rican endemic even though the species is also known from the U.S. Virgin Islands. The larger numbers associated with the bold lines connecting two

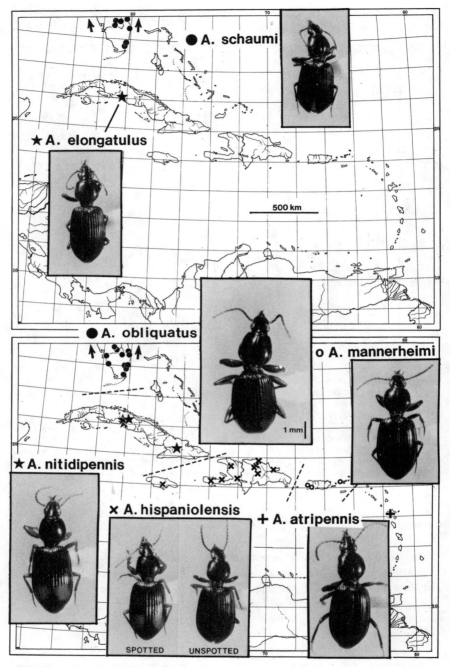

Figure 5-22. West Indian *Ardistomis* II: *A. schaumi* (specimen from 3.2 km northeast of Moncrief, Ga.; Cornell Univ. Insect Collections, Ithaca, N.Y.); *A. elongatulus* (specimen from Soledad nr. Cienfuegos, Cuba; Museum of Comparative Zoology, Harvard Univ., Cambridge

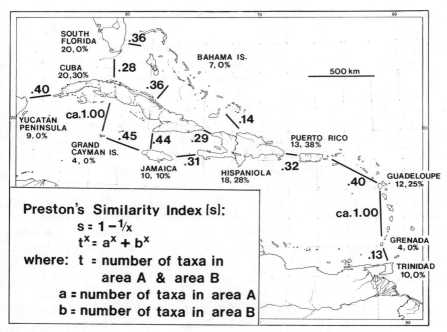

Figure 5-23. General summary of number of species, percent endemism, and similarity among selected areas in the West Indian Biogeographic Region. The number under each area name refers to the number of species recorded from that area. The percentage that follows this number refers to the percentage of the area's fauna that is endemic. The larger numbers associated with lines connecting two areas are Preston's Similarity Indices calculated using the formula presented in the inset at the bottom left. See the text for further information about this index.

land areas are Preston's Similarity Indices (s) (Preston 1962), calculated using the equation presented in the inset of Figure 5-23. This index and Preston's Dissimilarity Index ($z = 1 - s$) are measures of resemblance of two faunas that take into account the variations in the total number of species in the two regions being compared. Preston's Similarity Index ranges from zero to nearly (but not equaling) one. A value of zero means that two areas share no taxa in common. A value near one means that the fauna of one area (A) is the same or a subset of another area (B), that is, there is no taxon in area A that is not also in area B. The equation is transcendental and must be solved

---

Mass.); *A. obliquatus* (specimen from Mount Vernon, Ala.; Cornell Univ. Insect Collections, Ithaca); *A. mannerheimi* (specimen from Saint Thomas, Virgin Islands; Museum für Naturkunde der Humboldt-Universität, Berlin); *A. nitidipennis* (specimen from Soledad nr. Cienfuegos, Cuba; U.S. National Museum, Washington, D.C.); *A. hispaniolensis* (manuscript name) (spotted specimen from Ennery, Haiti, and unspotted specimen from San José de las Matas, Dominican Republic; both in Museum of Comparative Zoology, Harvard Univ.); and *A. atripennis* (specimen from Gourbeyre, Guadeloupe; American Museum of Natural History, New York City). All specimens photographed at the same scale.

either by trial and error (preferably with a computer) or using the table in Preston (1962:419). Williamson (1981:22–25) and Munroe (1984: 291) discuss this index. Only a sample of possible area comparisons was made. The island of Trinidad, although not considered part of the West Indian biogeographic region, was added for comparative purposes. Table 5-3 provides a checklist of the known scaritine fauna of Trinidad.

Cuba, the largest island in the West Indies, possesses the largest scaritine fauna (20 species) among islands. Other Caribbean islands generally follow a linear (log area)/(species number) relationship; however, Guadeloupe possesses a fauna larger than predicted based upon the island's size. The faunas of south Florida and Puerto Rico are the best sampled, making data for these two areas the most complete.

Of the 65 species recorded from the West Indian biogeographic region (as defined here), 27 (41.5%) are considered endemic. Fifty-three species are recorded from the West Indian islands, with 50.9% being endemic. The West Indian endemics are found on six islands: Cuba, Hispaniola, Jamaica, Puerto Rico, Guadeloupe, and Martinique. South Florida, the Yucatán Peninsula (Yucatán and Quintana Roo), the Bahama Islands, the Cayman Islands, Barbados, most of the Lesser Antillean islands, and Trinidad lack endemic species, probably because of their often low relief and their relatively dry climates and short lives as emergent land masses. Puerto Rico possesses the highest percent endemism (38%); however, future collecting on Cuba, Hispaniola, and Jamaica could raise the percent endemism of these islands through the discovery of undescribed, flightless species. If percent endemism can be regarded as an indication of how old an island is, then Puerto Rico ought to be judged the oldest island among the Greater Antilles. Values of Preston's Similarity Index relating Puerto Rico to Hispaniola and to Guadeloupe do not indicate that Puerto Rico is particularly isolated. The presence of three species of *Antilliscaris* on Puerto Rico also seems to point to the relative antiquity of this island. Jamaica ought to be judged relatively young

Table 5-3. Checklist of Trinidadian Scaritinae (Coleoptera: Carabidae)

Scarites (*Scarites*) *heterogrammus* Perty
S. (*Distichus*) *orientalis* Bonelli
Clivina (*Semiclivina*) *elongata* Chaudoir
C. (*Semiclivina*) sp.
Nyctosyles planicollis (Reiche)
Oxydrepanus micans Putzeys
Aspidoglossa schach (Fabricius)
A. striatipennis (Gory)
Semiardistomis pallipes (Dejean)
Ardistomis soror (Putzeys)

in comparison to Cuba, Hispaniola, and Puerto Rico. Buskirk (1985) summarized both geologic and biological data that support this view. Shields (1976) summarized the oldest ages recorded for rocks of the world's islands; however, because of compounding factors such as uplift, subsidence, volcanism, erosion, and sea-level changes, this information is not very good for estimating how long an island has been continuously above sea level or where the island was latitudinally and longitudinally when it first emerged.

Based on Preston's Similarity Index, Grand Cayman Island possesses a fauna that is a subset of the Cuban fauna, and therefore the faunas differ only in size. The fauna of Grenada is a subset of the Guadeloupe fauna—that is, the four species recorded from Grenada also occur on Gaudeloupe. The fauna of Grenada is very different from that of Trinidad, most likely because of the rapid transition from subtropical (Grenada) to tropical (Trinidad) climates at sea level; the fauna of Trinidad is dominated by widespread tropical species. The remaining comparisons (with similarity indexes varying from .45 to .28) show relatively high levels of isolation. The Bahama Islands taken together seem to be a biogeographic composite. If one includes the Bimini Islands and Grand Bahama Island with south Florida, the remaining Bahamian fauna becomes a subset of the Cuban fauna; also the similarity index relating the Bahama Islands to Hispaniola would rise from .14 to .20. None of the similarity values calculated approach the equilibium value, given by Preston as .73, implying that the West Indian scaritine fauna is in a state of flux. As suggested by Preston (1962) this flux might be due to the relative recency (18,000 years ago) of the Wisconsin glacial maximum that greatly lowered sea levels and produced drier climates in the West Indies (Pregill and Olson 1981).

Tables 5-4 and 5-5 provide ecological categorizations (by station) of

Table 5-4. Ecological stations of endemic West Indian Scaritinae

HALOPHOBIC HYGROBIONTS	1 riparian *Schizogenius*
	1 riparian *Semiardistomis*
	2 riparian and swamp-inhabiting *Clivina*, subg. *Semiclivina*
	5 swamp-inhabiting *Ardistomis*
	6 humicolous *Ardistomis*
	1 humicolous *Semiardistomis*
	4 humicolous ? *Antilliscaris*
	1 humicolous ? *Scarites*, subg. *Taeniolobus*
Total	23
HALOPHOBIC HYGROPHILES	1 *Scarites*, subg. *Scarites*
	1(2) *Clivina*, subg. *Paraclivina**
	1 *Clivina*, subg. *Isoclivina*
Total	3(4)
ECOLOGY UNKNOWN	1 *Stratiotes*

* *Clivina* nr. *latiuscula* may prove to be an endemic species.

Table 5-5. Ecological stations of nonendemic West Indian Scaritinae

HALOBIONTS		1 intertidal *Scarites*
		2 intertidal *Dyschirius*
		1 intertidal *Halocoryza*
	Total	4
HALOPHOBIC HYGROBIONTS		1 riparian *Clivina*, subg. *Reichardtula*
		1 riparian *Semiardistomis*
		3 riparian and swamp-inhabiting *Clivina*, subg. *Semiclivina*
		2 riparian *Dyschirius*
		1 swamp-inhabiting *Clivina*, subg. *Reichardtula*
		2 swamp-inhabiting *Oxydrepanus*
		2 swamp-inhabiting *Ardistomis*
		1 humicolous *Semiardistomis*
	Total	13
HALOPHOBIC HYGROPHILES		3 *Scarites*, subg. *Scarites*
		2 *Scarites*, subg. *Distichus*
		6(5) *Clivina*, subg. *Paraclivina**
		4 *Aspidoglossa*
	Total	15
XEROPHILES	Total	4 *Pasimachus*
ECOLOGY UNKNOWN		1 *Camptodontus*
		1 *Nyctosyles*
	Total	2

* *Clivina* nr. *latiuscula* may prove to be an endemic species.

endemic and nonendemic West Indian scaritines, respectively. The endemic West Indian scaritine fauna (Table 5-4) is dominated by halophobic hygrobionts (23 of 26 species for which the station is known or 88.5%). There are no endemic halobionts or xerophiles. Among the endemic species, riparian species and halophobic hygrophiles are the most widely distributed, occurring either on two or more islands or restricted to the largest island in the Caribbean—Cuba. These species are fully winged. Swamp-inhabiting species, also fully winged, are usually more restricted in their distributions. And flightless humicoles are even more localized with species frequently being restricted to single mountain ranges.

The nonendemic West Indian Scaritinae (Table 5-5) are not dominated by halophobic hygrobionts; only 36.1% (13 of 36 species) of the nonendemic West Indian scaritines are classified under this station. There are four nonendemic halobionts and four nonendemic xerophiles. The best represented station includes halophobic hygrophiles, species eurytopic with respect to their moisture requirements. Among the nonendemic halophobic hygrobionts, 61.5% (8 of 13 species) are species of the eastern United States that do not cross the Straits of Florida, thus contributing substantially to the dissimilarity between the faunas of south Florida and Cuba.

The preponderance of halophobic hygrobionts, species stenotopic with respect to moisture, among the endemic West Indian Scaritinae suggests that

Table 5-6. Summary by family of the number of endemic genera of West Indian Coleoptera (Insecta)

Family	No. endemic genera
Cerambycidae	43 (21 Lamiinae)
Chrysomelidae	25 (10 Alticinae)
Curculionidae	21 (7 Brachyderinae)
Pselaphidae	20 (15 Euplectini)
Staphylinidae	15 (7 Aleocharinae)
Tenebrionidae (incl. Alleculidae)	14 (8 Tenebrioniae)
Scarabaeidae	8 (5 Dynastinae)
Coccinellidae	7 (all Scymninae)
Lampyridae	5 (3 Photinini)
Buprestidae	4 (2 Buprestini)
Elmidae	4 (3 Elminae)
Anthribidae	2 (both Choraginae)
Brentidae	2
Colydiidae	2
Lycidae	2
Carabidae	2
Anobiidae, Anthicidae (incl. Pedilidae), Cleridae, Corylophidae, Cucujidae, Hemipeplidae, Monommidae, Phalacridae, Trogositidae (9 families)	1 each
Total	187

NOTE: Families after Hollis 1980.

fluctuating climates have played a prominent role in shaping the distributions and speciation patterns of West Indian fauna. Interestingly, riparian scaritines, well-represented at lower elevations, are virtually absent above one thousand m elevation. My best guess at an explanation is that riparian habitats at high elevations are extremely unstable because of the high gradience of the streams and the rainfall patterns that typically consist of midafternoon deluges in the rainy season.

Discussion

It is popular today to formulate historical biogeographic hypotheses based on explicit hypotheses of relationships among organisms, derived from cladistic analyses. By employing the methods of vicariance biogeography (Nelson and Platnick 1981) one can construct area cladograms that basically eliminate the need or desirability of elaborate and prolonged scenarios. Most of our current knowledge of the West Indian Scaritinae is at both the alpha-systematic and alpha-biogeographic level (i.e., the "descriptive level" of Ball 1975). It would therefore be nonsense to try to address beta- and gamma-level questions with the present information.

I am not optimistic that cladistic analyses of various insect groups will be useful in deciphering historical area relationships in the Caribbean in a geologic sense, as attempted by Rosen (1975, 1985). The geologic and climatic history of the Caribbean is apparently very complex. The West Indies have been colonized many times by insect groups exhibiting incredible diversity in vagility and ecology. Extinction and significant range changes have undoubtedly occurred among the West Indian Scaritinae; however, there are no fossil scaritines known from the region. From his studies of Dominican amber, Wilson (1985) noted significant changes in the ant fauna of Hispaniola since the late Oligocene or early Miocene. As noted previously, even today's West Indian scaritine fauna seems to be in a state of flux. Lowland areas (relatively poor in endemic species) appear to have been only recently colonized. In addition, there are a significant number of probable introductions by man.

Since the discovery of plate tectonics and the subsequent confirmation of continental drift, publications reporting biogeographic ties between the Afrotropical region and the Neotropical region have increased at a growing rate. Although there have been relatively few studies noting biogeographic ties between the West Indies and the Afrotropical region (and other parts of the Old World), it is not unreasonable to expect there to be some ties regardless of whether one thinks the West Indies are oceanic islands or fragments of a continental land mass. Some examples of insect groups in which faunal ties between the West Indies and the Old World (especially the Afrotropical region) have been noted include Scarabaeinae (Coleoptera, Scarabaeidae; Matthews 1966, Simonis 1981), Dynastinae (Coleoptera, Scarabaeidae; Howden 1970), Trichoptera and Odonata (Flint 1978), Rhopalocera (butterflies) (Lepidoptera; Brown 1978, Shields and Dvorak 1979), Carabidae (Coleoptera; Erwin 1979, Liebherr 1986), Orthocladiinae (Diptera, Chironomidae; Saether 1981), Steganinae (Diptera, Drosophilidae; see Grimaldi, chap. 8), and Lygaeidae (Heteroptera; see Slater, chap. 3).

Since the final separation of South America from Africa about 80 million years ago (Hallam 1981), there has been a considerable time for lineages to evolve, go extinct, or to become relictual in areas like the West Indies or Madagascar. I view *Antilliscaris* as one of these relict groups. I think that it is unreasonable to expect the distributions of organisms to mirror perfectly the historical geology of the earth, particularly since it is generally accepted that most species that have ever existed are extinct (Raup 1986). Unfortunately, the known fossil record of insects is pathetically poor, especially from subtropical and tropical regions.

Biogeographic relationships between the West Indies and the Old World will be most readily discovered by studying taxa at the generic level or higher that are endemic to the West Indies and attempting to relate these taxa to Old World groups as well as to New World groups. It is reasonable to hypothesize that old biogeographic relationships will be manifest at a relatively high taxonomic level. In addition, it may be easier to construct well-supported cladistic hypotheses by avoiding sexually selected and frequently autapomorphic or multistate characters that seem to plague species-level analyses.

I have compiled a list (available on request) of endemic genera of Coleoptera of the West Indies with the help of specialists who are listed in the acknowledgments. The total number of endemic genera is 187. Table 5-6 provides a summary by family. Certainly there are many systematic problems yet to be solved, and some of the genera may later prove invalid. Of the 159 families of Coleoptera recognized by Hollis (1980), 25 possess genera endemic to the West Indies. Three families (Cerambycidae, Chrysomelidae, and Curculionidae) include nearly half (89) of the endemic genera.

Assuming that the endemic genera represent the oldest element in the West Indian Coleoptera fauna (there are no endemic families or subfamilies), what preliminary observations are possible? First, the family with the largest number of endemic genera (Cerambycidae) possesses wood-boring larvae. Second, nearly all the remaining families include members with larvae that bore in wood, twigs, stems, fruits, or roots, and they are also commonly encountered under bark. Although a much finer level of analysis is warranted, I think these data are consistent with the view that the West Indian islands (including the Greater Antilles) emerged as oceanic islands and that many of the early terrestrial colonizers arrived by rafting.

Summary

Sixty-five species belonging to the subfamily Scaritinae are known from the West Indian biogeographic region (including south Florida and part of the Yucatán Peninsula). Of these, 27 (41.5%) are judged to be endemic to the region. Single-island endemics are found on six islands: Cuba, Hispaniola, Jamaica, Puerto Rico, Guadeloupe, and Martinique. South Florida, the Yucatán Peninsula, the Bahama Islands, the Cayman Islands, Barbados, Trinidad, and most of the Lesser Antillean islands lack endemic species, probably due to their often low relief (and associated dry climates) and relatively short lives as emergent land masses.

Antilliscaris (known from Puerto Rico and Hispaniola) is one of two endemic genera of West Indian Carabidae. Based on larval characters, this genus is closely aligned with genera of the Afrotropical region. The presence of *Antilliscaris* (3 species) and the relatively high percentage of endemism (38%) in the Puerto Rican scaritine fauna suggest that Puerto Rico is the oldest emergent land mass in the West Indies.

The endemic West Indian scaritine fauna is dominated by *halophobic hygrobionts* (salt-intolerant species that require high levels of moisture), suggesting that fluctuating climates have played a prominant role in shaping the West Indian fauna. Based on area comparisons employing Preston's Similarity Index, the West Indian scaritine fauna appears to be in a state of flux, perhaps due to eustatic and climatic perturbations in the Pleistocene and during the last glacial maximum 18,000 years ago. In addition, there are significant numbers of probable introductions by man, particularly among *halophobic hygrophiles* (salt-intolerant species that are eurytopic with respect to their moisture requirements).

The distribution of *Pasimachus* and a preliminary compilation of endemic genera of West Indian Coleoptera support the view that the Greater Antilles are oceanic islands, first colonized by overwater dispersal from the mainland, especially by rafting.

Acknowledgments

I thank James Liebherr for his invitation to participate in the symposium that led to this volume. I gratefully acknowledge all the institutions and persons that made material available for this study. These will be individually noted in a taxonomic paper on the West Indian Scaritinae. I thank the National Science Foundation (Washington, D.C.) for enabling me to examine type material in several European repositories through a Doctoral Dissertation Improvement Grant (BSR83–10867). I am grateful for the help and patience of the following coleopterists who aided me in compiling a list of endemic West Indian genera of Coleoptera: D. S. Chandler, J. A. Chemsak, R. D. Gordon, A. T. Howden, H. F. Howden, C. W. O'Brien, T. J. Spilman, R. E. White, D. R. Whitehead, and S. L. Wood. I take full responsibility for any errors introduced into this list. I thank the following individuals for reviewing the manuscript: G. C. Eickwort (Cornell University), T. L. Erwin (U.S. National Museum of Natural History), J. H. Frank (University of Florida), H. F. Howden (Carleton University), J. K. Liebherr (Cornell University), S. B. Peck (Carleton University), and Q. D. Wheeler (Cornell University). Finally, I acknowledge F. Robert Wesley (Cornell University) for his expert help in photographing the insects in Figures 5-3 through 5-18.

References

Ball, I.R. 1975. Nature and formulation of biogeographic hypotheses. Syst. Zool. 24:407–430.

Bänninger, M. 1937. Über südamerikanische *Taeniolobus*—Arten (Col. Carab.). Entomol. Bl. Biol. Syst. Kaefer 33:320–322.

———. 1938. Monographie der subtribus Scaritina (Col. Carab.). II. Dtsch. Entomol. Z. 1938:41–181, Taf. 1–4.

————. 1941. Bestimmungstabelle der südamerikanischen Formen des subg. *Taeniolobus* Chd. (Col. Carab.). Entomol. Bl. Biol. Syst. Kaefer 37:67–78.

————. 1950. The subtribus Pasimachina (Coleoptera, Carabidae, Scaritini). Revista de Entomol. 21:481–511.

Basilewsky, P. 1973. Insectes Coléoptères. Carabidae Scaritinae. Faune Madagascar 37:1–322.

————. 1976. Insectes Coléoptères. Carabidae Scaritinae, III: Supplément à la systématique. Faune Madagascar 41:163–220.

Bates, H.W. 1881. Insecta, Coleoptera, Cicindelidae, Carabidae, vol. 1, pt. 1 (in part; pp. 1–40, pls. i–ii). *In* F.D. Godman and O. Salvin, eds., Biologia Centrali-Americana. London.

Beard, J.S. 1949. The natural vegetation of the Windward and Leeward islands. Oxford For. Mem. 21:1–192.

Blatchley, W.S. 1910. An illustrated and descriptive catalogue of the Coleoptera or beetles (exclusive of the Rhynchophora) known to occur in Indiana. Nature, Indianapolis. 1385 pp.

Briggs, J.C. 1984. Freshwater fishes and biogeography of Central America and the Antilles. Syst. Zool. 33:428–435.

Brown, F.M. 1978. The origins of the West Indian butterfly fauna. *In* F.B. Gill, ed., Zoogeography in the Caribbean, pp. 5–30. The 1975 Leidy Medal Symposium. Special Publication No. 13, Academy of Natural Sciences of Philadelphia.

Bruneau de Miré, P. 1979. Trans-Atlantic dispersal: Several examples of the colonization of the Gulf of Biafra by Middle American stocks of Carabidae. *In* T.L. Erwin, G.E. Ball, and D.R. Whitehead, eds., Carabid Beetles: Their Evolution, Natural History, and Classification, pp. 327–330. Dr. W. Junk bv Publishers, The Hague.

Buskirk, R.E. 1985. Zoogeographic patterns and tectonic history of Jamaica and the northern Caribbean. J. Biogeogr. 12:445–461.

Candolle, A.-P. de. 1820. Géographie botanique. *In* Dictionnaire des Sciences Naturelles, vol. 18. F.G. Levrault, Strasbourg.

Darlington, P.J., Jr. 1934. New West Indian Carabidae, with a list of Cuban species. Psyche 41:66–131.

————. 1935. Three West Indian Carabidae in Florida. Psyche 42:161.

————. 1936. West Indian Carabidae, II: Itinerary of 1934; forests of Haiti; new species; and a new key to *Colpodes*. Psyche 42:167–215.

————. 1937. West Indian Carabidae, III: New species and records from Cuba, with a brief discussion of the mountain fauna. Mem. Soc. Cubana Hist. Nat. 11:115–136.

————. 1938. The origin of the fauna of the Greater Antilles, with discussion of dispersal of animals over water and through the air. Quart. Rev. Biol. 13:274–300.

————. 1939. West Indian Carabidae, V: New forms from the Dominican Republic and Puerto Rico. Mem. Soc. Cubana Hist. Nat. 13:79–101.

————. 1941. West Indian Carabidae, VI: The Jamaican species and their wings. Psyche 48:10–15.

————. 1943. Carabidae of mountains and islands: Data on the evolution of isolated faunas and on the atrophy of wings. Ecol. Monogr. 13:37–61.

————. 1947. West Indian Carabidae (Col.), VII: The species of the Cayman Islands. Entomol. Mon. Mag. 83:209–211.

————. 1953a. West Indian Carabidae (Coleoptera), the Bahama species. Amer. Mus. Novit. 1650:1–16.

————. 1953b. West Indian Carabidae, IX: More about Jamaican species. Occ. Pap. Mus. Inst. Jam. 8:1–14.

————. 1970a. Coleoptera: Carabidae including Cicindelinae. Insects Micronesia 15:1–49.

————. 1970b. Carabidae of tropical islands, especially the West Indies. Biotropica 2:7–15.

————. 1971. The carabid beetles of New Guinea. Part IV: General considerations; analysis and history of fauna; taxonomic supplement. Bull. Mus. Comp. Zool. Harv. 142:129–337.

den Boer, P.J. 1985. Exclusion, competition or coexistence? A question of testing the right hypotheses. Z. Zool. Syst. Evolutionsforsch. 23:259–274.

Diamond, J.M. 1985. Rats as agents of extermination. Nature 318:602–603.

Dunn, G.A. 1986. Beetle collecting in Jamaica, with special notes on the tiger beetles. Young Entomol. Soc. Quart. 3:8–13.

Emden, F.I. van. 1942. A key to the genera of larval Carabidae (Col.). Trans. R. Entomol. Soc. Lond. 92:1–99.

Endler, J.A. 1982. Alternative hypotheses in biogeography: Introduction and synopsis of the symposium. Amer. Zool. 22:349–354.

Erwin, T.L. 1978. The larva of *Enceladus gigas* Bonelli (Coleoptera: Carabidae: Siagoninae: Enceladini) with notes on the phylogeny and classification of some more primitive tribes of ground beetles. Coleopt. Bull. 32:99–106.

————. 1979. The American connection, past and present, as a model blending dispersal and vicariance in the study of biogeography. *In* T.L. Erwin, G.E. Ball, and D.R. Whitehead, eds., Carabid Beetles: Their Evolution, Natural History, and Classification, pp. 355–367. Dr. W. Junk bv Publishers, The Hague.

————. 1981. Natural history of Plummers Island, Maryland, XXVI: The ground beetles of a temperate forest site (Coleoptera: Carabidae): An analysis of fauna in relation to size, habitat selection, vagility, seasonality, and extinction. Bull. Biol. Soc. Wash. 5:105–224.

————. 1982. Small terrestrial ground beetles of Central America (Carabidae: Bembidiina and Anillina). Proc. Calif. Acad. Sci. 42:455–496.

Erwin, T.L., and L.L. Sims. 1984. Carabid beetles of the West Indies (Insecta: Coleoptera): A synopsis of the genera and checklists of the tribes of Caraboidea, and of the West Indian species. Quaest. Entomol. 20:351–466.

Ewel, J.J., and J.L. Whitmore. 1973. The ecological life zones of Puerto Rico and the U.S. Virgin Islands. U.S. Forest Service, Forest Research Paper ITF-18. 72 pp., 1 map.

Ewel, J.J., A. Madriz, and J.A. Tosi, Jr. 1968. Zonas de vida de Venezuela; y mapa ecológico de Venezuela. Ministerio de Agricultura y Cría, Dirección de Investigación, Caracas. 265 pp., 1 map.

Fleutiaux, E., and A. Sallé. 1889. Liste des Coléoptères de la Guadeloupe et descriptions d'espèces nouvelles. Ann. Soc. Entomol. France (6)9:351–484.

Flint, O.S., Jr. 1978. Probable origins of the West Indian Trichoptera and Odonata faunas. *In* M.I. Crichton, ed., Proceedings of the 2d International Symposium on

Trichoptera, pp. 215–223. University of Reading, England, 25–29 July 1977. Dr. W. Junk bv Publishers, The Hague.

Granger, O.E. 1985. Caribbean climates. Progr. Phys. Geogr. 9:16–43.

Gray, K.M., and G.A. Symes. 1969. Forestry Inventory Map, no. 3, Life Zones, 1:500,000. United Nations Development Program, Forestry/Management Project, Kingston, Jamaica.

Hallam, A. 1981. Relative importance of plate movements, eustasy, and climate in controlling major biogeographical changes since the early Mesozoic. *In* G. Nelson and D.E. Rosen, eds., Vicariance Biogeography: A Critique, pp. 303–330. Columbia University Press, New York.

Hennig, W. 1966. Phylogenetic Systematics. University of Illinois Press, Urbana. 268 pp.

Hlavac, T.F. 1969. A review of the species of *Scarites* (*Antilliscaris*) (Coleoptera: Carabidae) with notes on their morphology and evolution. Psyche 76:1–17.

Hollis, D., ed. 1980. Animal Identification: A Reference Guide, vol. 3: Insecta. British Museum (Natural History), London, and John Wiley and Sons, Ltd., Chichester, Sussex. viii, 160 pp.

Howden, H.F. 1970. Jamaican Scarabaeidae: Notes and descriptions (Coleoptera). Can. Entomol. 102:1–15.

Kryzhanovsky, O.L. 1976. Revised classification of the family Carabidae. Entomol. Rev. (Engl. Transl. Entomol. Obozr.) 55:80–91.

Kult, K. 1947. The 3rd study to the knowledge of tribus Clivinini (Col. Carab.). Čas. Česk. Spol. Entomol. 44:26–37.

———. 1959. Revision of the African species of the old genus *Clivina* Latr. (Col. Carabidae). Rev. Zool. Bot. Afr. 60:172–225.

Liebherr, J.K. 1986. *Barylaus*, new genus (Coleoptera: Carabidae) endemic to the West Indies with Old World affinities. J. N.Y. Entomol. Soc. 94:83–97.

Lindroth, C.H. 1961. The ground beetles (Carabidae, excl. Cicindelinae) of Canada and Alaska, part. 2. Opusc. Entomol. Suppl. 20:1–200.

MacLean, W.P., and R.D. Holt. 1979. Distributional patterns in St. Croix *Sphaerodactylus* lizards: The taxon cycle in action. Biotropica 11:189–195.

Matthews, E.G. 1966. A taxonomic and zoogeographic survey of the Scarabaeinae of the Antilles (Coleoptera: Scarabaeidae). Mem. Amer. Entomol. Soc. 21:1–134.

Munroe, E.G. 1984. Biogeography and evolutionary history: Wide-scale and long-term patterns of insects. *In* C.B. Huffaker and R.L. Rabb, eds., Ecological Entomology, pp. 279–304. John Wiley and Sons, New York.

Myers, R.L. 1986. Florida's freezes: An analog of short-duration nuclear winter events in the tropics. Florida Sci. 49:104–115.

Nelson, G. 1978. From Candolle to Croizat: Comments on the history of biogeography. J. Hist. Biol. 11:269–305.

Nelson, G., and N. Platnick. 1981. Systematics and Biogeography: Cladistics and Vicariance. Columbia University Press, New York. 567 pp.

Nichols, S.W. 1985. *Clivina* (*Semiclivina*) *vespertina* Putzeys, a probable introduction to the United States from South America. Coleopt. Bull. 39:380.

———. 1986a. Two new flightless species of *Scarites* s. str. inhabiting Florida and the West Indies (Coleoptera: Carabidae: Scaritini). Proc. Entomol. Soc. Wash. 88:257–264.

———. 1986b. Descriptions of larvae of Puerto Rican species of *Antilliscaris* Bänninger and notes about relationships and classification of *Antilliscaris* (Coleoptera: Carabidae: Scaritini: Scaritina). Coleopt. Bull. 40:301–311.

Nickle, D.A., and J.L. Castner 1984. Introduced species of mole crickets in the United States, Puerto Rico, and the Virgin Islands (Orthoptera: Gryllotalpidae). Ann. Entomol. Soc. Amer. 77:450–465.

OAS (Organization of American States). 1972. Ecologie, Republique d'Haiti. Washington, D.C.

Peck, S.B. 1975. The invertebrate fauna of tropical American caves. 3: Jamaica, an introduction. Int. J. Speleol. 7:303–326.

Pregill, G.K., and S.L. Olson. 1981. Zoogeography of West Indian vertebrates in relation to Pleistocene climatic cycles. Ann. Rev. Ecol. Syst. 12:75–98.

Preston, F.W. 1962. The canonical distribution of commonness and rarity. Ecology 43:185–215, 410–432.

Raup, D.M. 1986. Biological extinction in earth history. Science 231:1528–1533.

Reichardt, H. 1977. A synopsis of the genera of Neotropical Carabidae (Insecta: Coleoptera). Quaest. Entomol. 13:346–493.

Rosen, D.E. 1975. A vicariance model of Caribbean biogeography. Syst. Zool. 24:431–464.

———. 1985. Geological hierarchies and biogeographic congruence in the Caribbean. Ann. Missouri Bot. Gard. 72:636–659.

Saether, E.A. 1981. Orthocladiinae (Chironomidae: Diptera) from the British West Indies with descriptions of *Antillocladius* n. gen., *Lipurometricnemus* n. gen., *Competrosmittia* n. gen., and *Diplosmittia* n. gen. Entomol. Scand. 16:1–46.

Shields, O. 1976. A summary of the oldest ages for the world's islands. Pap. Proc. R. Soc. Tasmania 110:35–61.

Shields, O., and S.K. Dvorak. 1979. Butterfly distribution and continental drift between the Americas, the Caribbean and Africa. J. Nat. Hist. 13:221–250.

Simonis, A. 1981. *Anoplodrepanus*, nuovo genere di Oniticellini (Coleoptera, Scarabaeidae). Boll. Mus. Zool. Univ. Torino 7:87–94.

Stephens, J.F. 1827. Illustrations of British Entomology; or a synopsis of indigenous insects: Containing their generic and specific distinctions; with an account of their metamorphoses, times of appearance, localities, food, and ecology, as far as practicable. Mandibulata. Vol. I (in part), pp. 1–76, pls. 1–5.

Tasaico, H. 1967. Reconocimiento y evaluación de los recursos naturales de la República Dominicana: Estudio para su desarrollo y planificación (con Mapa Ecológico de la República Dominicana, 1:250,000). Pan American Union, Washington, D.C.

Thiele, H.-U. 1977. Carabid Beetles in Their Environments: A Study on Habitat Selection by Adaptations in Physiology and Behavior. Springer-Verlag, New York. 369 pp.

Whitehead, D.R. 1966. A review of *Halocoryza* Alluaud, with notes on its relationship to *Schizogenius* Putzeys (Coleoptera: Carabidae). Psyche 73:217–228.

———. 1969a. Variation and distribution of the intertidal beetle *Halocoryza arenaria* (Darlington) in Mexico and the United States (Coleoptera: Carabidae). J. N.Y. Entomol. Soc. 77:36–39.

————. 1969b. Notes on *Dyschirius* Bonelli and *Akephorus* LeConte, with a peculiar new *Dyschirius* from Texas (Coleoptera: Carabidae: Scaritini). J. N.Y. Entomol. Soc. 77:179–192.

————. 1972. Classification, phylogeny, and zoogeography of *Schizogenius* Putzeys (Coleoptera: Carabidae: Scaritini). Quaest. Entomol. 8:131–348.

————. 1974. *Clivina texana* LeConte, a synonym of *C. planicollis* LeConte (Coleoptera: Carabidae: Scaritini). Proc. Entomol. Soc. Wash. 76:454.

————. 1979. Speciation patterns and what they mean. *In* T.L. Erwin, G.E. Ball, and D.R. Whitehead, eds., Carabid Beetles: Their Evolution, Natural History, and Classification, pp. 23–33. Dr. W. Junk bv Publishers, The Hague.

Williamson, M. 1981. Island Populations. Oxford University Press, Oxford. xi, 286 pp.

Wilson, E.O. 1985. Invasion and extinction in the West Indian ant fauna: Evidence from the Dominican amber. Science 229:265–267.

Appendix: Distributional Checklist of West Indian Scaritinae

Column groups: **Greater Antilles** = Cuba, Isla de Pinos, Cayman Islands, Jamaica, Hispaniola, Puerto Rico. **Leeward Islands** = St. Thomas, St. John, St. Croix, Guadeloupe-Basse Terre, Guadeloupe-Grande Terre, Dominica. **Windward Islands** = Martinique, St. Lucia, St. Vincent, Mustique Island, Grenada, Barbados. (Columns under heading "West Indian Biogeographic Region.")

Taxon	South Florida	Yucatán Peninsula	Bahama Islands	Cuba	Isla de Pinos	Cayman Islands	Jamaica	Hispaniola	Puerto Rico	St. Thomas	St. John	St. Croix	Guadeloupe-Basse Terre	Guadeloupe-Grande Terre	Dominica	Martinique	St. Lucia	St. Vincent	Mustique Island	Grenada	Barbados	Eastern U.S.A. (excl. south Florida)	Mexico (excl. Yucatán Peninsula)	South America
Tribe PASIMACHINI (New Status)																								
1. *Pasimachus (Pasimachus) floridanus* Casey 1913:79	+	−	−	−	−	−	−	−	−	−	−	−	−	−	−	−	−	−	−	−	−	+	−	−
2. *Pasimachus (Pasimachus) marginatus* (Fabricius 1775:94)	+	−	−	−	−	−	−	−	−	−	−	−	−	−	−	−	−	−	−	−	−	+	−	−
3. *Pasimachus (Pasimachus) sublaevis* (Palisot de Beauvois 1805–21:107)	+	−	−	−	−	−	−	−	−	−	−	−	−	−	−	−	−	−	−	−	−	+	−	−
4. *Pasimachus (Emydopterus) purpuratus* (Putzeys 1845:362)	−	+	−	−	−	−	−	−	−	−	−	−	−	−	−	−	−	−	−	−	−	−	+	−
Tribe DYSCHIRIINI																								
5. *Dyschirius (Dyschirius) abbreviatus* Putzeys 1846:532	+	−	−	−	−	−	−	−	−	−	−	−	−	−	−	−	−	−	−	−	−	+	−	−
6. *Dyschirius (Dyschirius)* nr. *erythrocerus* LeConte 1857:78	+	+	+	+	−	−	−	+	−	−	−	−	−	−	−	−	−	−	−	−	−	+	−	−
7. *Dyschirius (Dyschirius) pumilus* (Dejean 1825:425)	+	−	−	−	−	−	−	−	−	−	−	−	−	−	−	−	−	−	−	−	−	+	−	−
8. *Dyschirius (Dyschirius) sublaevis* Putzeys 1846:562	+	+	−	+	−	−	−	−	−	−	−	−	−	−	−	−	−	−	−	−	−	+	−	−
Tribe SCARITINI																								
9. *Scarites (Scarites) alternans* Chaudoir 1843:729	−	−	−	+	−	−	−	−	−	−	−	−	−	−	−	−	−	−	−	−	−	−	−	−
10. *Scarites (Scarites) illustris* Chaudoir 1880:66,91	−	−	−	+	−	−	−	−	−	−	−	−	−	−	−	−	−	−	−	−	−	−	+	+

(continued)

	11	12	13	14	15	16	17	18	19	20	21	22	23	24	25	26
	−	+	+	−	+	+	−	−	−	−	−	−	+	−	−	+
	+	−	−	−	+	−	−	−	−	−	−	−	−	−	−	−
	+	+	+	−	−	+	−	−	−	−	−	−	+	−	−	+
	−	−	−	−	−	−	−	−	−	−	−	−	−	−	−	−
	−	−	−	−	−	−	−	−	−	−	−	−	+	−	−	−
	−	−	−	−	−	−	−	−	−	−	−	−	+	−	−	−
	−	−	−	−	−	−	−	−	−	−	−	−	+	−	−	−
	−	−	−	−	−	−	−	−	−	−	−	+	+	−	−	−
	−	−	−	−	−	−	−	−	−	−	+	−	+	−	−	−
	−	−	−	−	−	−	−	−	−	−	−	−	+	−	−	−
	−	−	−	−	+	−	−	−	−	−	−	−	−	−	−	−
	−	−	−	−	−	−	−	−	−	−	−	−	+	−	−	−
	−	−	−	−	−	−	−	−	−	−	−	−	+	−	−	−
	−	−	−	+	−	−	+	−	+	+	−	−	−	−	−	−
	−	−	−	−	+	−	−	+	−	−	−	−	+	+	+	−
	−	−	+	−	−	+	−	+	−	−	−	−	+	−	+	−
	+	−	+	−	+	−	−	−	−	−	−	−	+	+	−	+
	+	−	−	−	+	−	−	−	−	−	−	−	+	+	−	−
	+	+	−	−	−	−	−	−	−	−	−	−	+	−	+	−

11. *Scarites* (*Scarites*) *marinus*
 Nichols 1986:258
12. *Scarites* (*Scarites*) *quadriceps*
 Chaudoir 1843:729
13. *Scarites* (*Scarites*) *subterraneus*
 Fabricius 1775:249
14. *Scarites* (*Taeniolobus*) *cubanus*
 Bänninger 1937:321
15. *Scarites* (*Distichus*) *octocoelus*
 (Chaudoir 1855:50)
16. *Scarites* (*Distichus*) *orientalis*
 Bonnelli 1813:469
17. *Antilliscaris danforthi*
 (Darlington 1939:80) (new combination)
18. *Antilliscaris darlingtoni*
 (Bänninger 1935:159) (new combination)
19. *Antilliscaris megacephalus*
 (Hlavac 1969:4) (new combination)
20. *Antilliscaris mutchleri*
 (Bänninger 1939:149) (new combination)

Tribe FORCIPATORINI
21. *Stratiotes iracundus*
 Putzeys 1863:9
22. *Camptodontus anglicanus*
 (Stephens 1827:38)

Tribe CLIVININI
23. *Halocoryza arenaria*
 (Darlington 1939:84)
24. *Schizogenius* (*Schizogenius*) *arimao*
 Darlington 1934:71
25. *Clivina* (*Isoclivina*) *limbipennis*
 Jacquelin Duval 1857:16
26. *Clivina* (*Reichardtula*) *americana*
 Dejean 1831:503

Appendix *(Continued)*

Grouping within the **West Indian Biogeographic Region**: Greater Antilles = Cuba–Puerto Rico; Leeward Islands = St. Thomas–Dominica; Windward Islands = Martinique–Barbados.

Taxon	South Florida	Yucatán Peninsula	Bahama Islands	Cuba	Isla de Pinos	Cayman Islands	Jamaica	Hispaniola	Puerto Rico	St. Thomas	St. John	St. Croix	Guadeloupe-Basse Terre	Guadeloupe-Grande Terre	Dominica	Martinique	St. Lucia	St. Vincent	Mustique Island	Grenada	Barbados	Eastern U.S.A. (excl. south Florida)	Mexico (excl. Yucatán Peninsula)	South America
Tribe CLIVININI (cont.)																								
27. *Clivina (Reichardtula) acuducta* Haldeman 1843a:296,299	+	–	–	–	–	–	–	–	–	–	–	–	–	–	–	–	–	–	–	–	–	+	–	–
28. *Clivina (Semiclivina) addita* Darlington 1934:67	–	–	–	–	–	–	–	+	+	–	–	–	–	–	–	–	–	–	–	–	–	–	–	–
29. *Clivina (Semiclivina) cubae* Darlington 1934:68	–	–	–	+	–	–	–	–	–	–	–	–	–	–	–	–	–	–	–	–	–	–	–	–
30. *Clivina (Semiclivina) dentipes* Dejean 1825:415	–	–	–	+	–	–	+	–	–	–	–	–	+	–	–	–	–	–	–	–	–	+	+	–
31. *Clivina (Semiclivina) elongata* Chaudoir 1843:734	–	–	–	–	–	–	–	–	–	–	–	–	–	–	–	–	–	–	–	–	–	–	–	+
32. *Clivina (Semiclivina) oblita* Putzeys 1866:168	–	–	–	–	+	–	–	–	–	–	–	–	–	–	–	–	+	–	–	–	–	–	–	+
33. *Clivina (Paraclivina) biguttata* Putzeys 1866:155,157	–	–	–	+	–	+	+	–	–	–	–	–	–	–	–	–	–	–	–	–	–	–	–	–
34. *Clivina (Paraclivina) bipustulata* (Fabricius 1798:44)	+	+	+	+	–	+	+	+	–	–	–	–	–	–	–	–	–	–	–	–	–	+	+	–
35. *Clivina (Paraclivina) fasciata* Putzeys 1846:624	–	–	–	+	–	–	–	+	+	+	–	+	+	–	+	–	–	–	–	+	–	+	+	+
36. *Clivina (Paraclivina) nr. latiuscula* Putzeys 1866:154	–	–	–	–	–	–	–	–	–	–	–	–	+	–	–	–	–	–	–	–	–	–	–	–
37. *Clivina (Paraclivina) marginipennis* Putzeys 1846:619	–	–	–	–	–	–	–	–	+	–	–	–	+	+	+	–	+	–	–	–	–	+	+	–
38. *Clivina (Paraclivina) tristis* Putzeys 1846:620	–	–	–	–	–	–	–	–	–	–	–	–	–	–	–	–	–	–	+	–	–	–	–	+

#	Species																								
39.	*Clivina (Paraclivina) tuberculata* Putzeys 1846:615	−	−	−	−	−	−	−	−	−	−	−	−	+	+	+	+	+	+	+	+	+	+	−	+
40.	*Nyctosyles planicollis* (Reiche 1842:376)	−	−	−	−	−	−	−	−	+	−	−	−	−	−	−	−	−	−	−	−	+	−	−	+
41.	*Oxydrepanus coamensis* (Mutchler 1934:2)	−	−	−	−	−	−	−	−	−	−	−	+	−	−	−	−	−	−	−	−	−	−	−	−
42.	*Oxydrepanus micans* Putzeys 1866:105	−	−	+	−	−	−	−	−	+	+	+	+	−	−	−	−	−	−	−	−	+	−	−	+
43.	*Oxydrepanus reicheoides* Darlington 1939:83	+	+	+	−	−	−	−	−	+	−	−	+	−	−	−	−	−	−	−	−	−	−	−	−
44.	*Oxydrepanus rufus* (Putzeys 1846:564)	+	+	+	+	−	−	−	−	−	−	−	−	−	−	−	−	−	−	−	+	−	+	−	+
45.	*Aspidoglossa cribrata* Putzeys 1846:634	−	+	−	+	−	−	+	+	+	+	+	−	+	+	+	+	+	+	+	+	+	−	+	−
46.	*Aspidoglossa mexicana* (Chaudoir 1837:18)	+	−	+	−	−	−	−	−	−	−	−	−	−	−	−	−	−	−	−	−	−	+	−	−
47.	*Aspidoglossa schach* (Fabricius 1792:153) (new combination)	+	+	+	+	−	−	−	+	+	+	+	+	+	+	+	+	+	+	+	+	+	−	+	+
48.	*Aspidoglossa subangulata* (Chaudoir 1843:738)	+	−	−	−	−	−	−	−	−	−	−	−	−	−	−	−	−	−	−	−	−	−	−	−
49.	*Semiardistomis cyaneolimbatus* (Chevrolat 1863:194) (new combination)	−	+	+	+	−	−	−	−	−	−	−	−	−	−	−	−	−	+	−	−	+	−	−	−
50.	*Semiardistomis laevistriatus* (Fleutiaux and Sallé 1889:363) (new combination)	−	−	−	−	−	−	−	−	−	+	−	−	−	−	−	−	−	−	−	−	−	−	−	−
51.	*Semiardistomis puncticollis* (Dejean 1831:508) (new combination)	+	−	+	−	−	−	−	−	−	−	−	−	−	−	−	−	−	−	−	−	+	−	+	−
52.	*Semiardistomis viridis* (Say 1825:21) (new combination)	+	+	−	−	−	−	−	−	−	−	+	+	+	−	−	−	−	−	−	−	+	−	+	−
53.	*Ardistomis alticolus* Darlington 1935b:173	−	−	−	−	−	−	−	−	−	+	−	−	−	−	−	−	−	−	−	−	−	−	−	−
54.	*Ardistomis atripennis* Putzeys 1866:202	−	−	−	−	−	−	−	−	+	−	−	−	−	−	−	−	−	−	+	−	−	−	−	−
55.	*Ardistomis elongatulus* Putzeys 1866:208	+	−	−	−	−	−	−	−	−	−	−	−	−	−	−	−	−	−	−	−	−	−	−	−

(continued)

Appendix (*Continued*)

Taxon	West Indian Biogeographic Region																					Eastern U.S.A. (excl. south Florida)	Mexico (excl. Yucatán Peninsula)	South America
	South Florida	Yucatán Peninsula	Bahama Islands	Cuba	Isla de Pinos	Cayman Islands	Jamaica	Hispaniola	Puerto Rico	St. Thomas	St. John	St. Croix	Guadeloupe-Basse Terre	Guadeloupe-Grande Terre	Dominica	Martinique	St. Lucia	St. Vincent	Mustique Island	Grenada	Barbados			
Tribe CLIVININI (cont.)																								
56. *Ardistomis franki* Nichols (manuscript name)	—	—	—	—	—	—	+	—	—	—	—	—	—	—	—	—	—	—	—	—	—	—	—	—
57. *Ardistomis guadeloupensis* Kult 1950:307	—	—	—	—	—	—	—	—	—	—	—	—	+	—	—	—	—	—	—	—	—	—	—	—
58. *Ardistomis hispaniolensis* Nichols (manuscript name)	—	—	—	—	—	—	+	+	—	—	—	—	—	—	—	—	—	—	—	—	—	—	—	—
59. *Ardistomis mannerheimi* Putzeys 1846:645	—	—	—	—	—	—	—	—	+	+	—	—	—	—	—	—	—	—	—	—	—	—	—	—
60. *Ardistomis nigroclarus* Darlington 1939:83	—	—	—	—	—	—	—	+	—	—	—	—	—	—	—	—	—	—	—	—	—	—	—	—
61. *Ardistomis nitidipennis* Darlington 1934:70	—	—	—	+	—	—	—	—	—	—	—	—	—	—	—	—	—	—	—	—	—	—	—	—
62. *Ardistomis obliquatus* Putzeys 1866:638	+	—	—	—	—	—	—	—	—	—	—	—	—	—	—	—	—	—	—	—	—	—	—	—
63. *Ardistomis ramsdeni* Darlington 1937:120	—	—	—	+	—	—	—	—	—	—	—	—	—	—	—	—	—	—	—	—	—	+	—	—
64. *Ardistomis rufoclarus* Darlington 1939:82	—	—	—	—	—	—	—	+	—	—	—	—	—	—	—	—	—	—	—	—	—	—	—	—
65. *Ardistomis schaumi* LeConte 1857:80	+	—	—	—	—	—	—	—	—	—	—	—	—	—	—	—	—	—	—	—	—	+	—	—

6 · Biogeographic Patterns of West Indian *Platynus* Carabid Beetles (Coleoptera)

James K. Liebherr

The *Platynus* beetle fauna of the West Indies comprises at least 66 species, most of which are restricted to single islands, and some to only one mountain on an island. This spectacularly high level of endemism is not atypical for the genus *Platynus*, which comprises more than 300 species in Mexico and Central America (Whitehead 1973; Fig. 6-1) and 1000 or more species worldwide. Many of these species are restricted to upland or montane habitats. The fragmented aspect of such habitats, coupled with frequent reduction of the metathoracic flight wings of the adult beetles, is correlated with proliferation of geographically restricted species on mainland and island areas alike.

Of the Greater Antilles, Hispaniola supports the largest number of *Platynus* species (Fig.6-1), with Jamaica second in diversity, far above the proportion expected by virtue of its area. Only 13 species are known from Cuba, possibly an artifact of limited collecting. However, as most *Platynus* species are found in habitats above 300 m elevation, Cuba may be considered an archipelago of islands surrounded by a terrestrial sea of inhospitable habitats. Puerto Rico possesses a central cordillera with substantial forested habitat, but no *Platynus* have been taken there. The geographic isolation of Puerto Rico may be cause for such lack of species, as suggested by findings reported in this paper.

The Lesser Antilles support a fauna of 11 endemic species, Guadeloupe and Dominica being the most diverse with 5 and 4 species, respectively.

A data set of 66 congeners upon which to base biogeographic hypotheses appears an ideal situation. Amazingly, before 1935 only 8 of these species were known, only one of these from the Greater Antilles. Based on three trips to the Greater Antilles in 1934, 1936, and 1938, Philip J. Darlington, Jr., collected material that enabled him to describe 47 species of *Platynus*

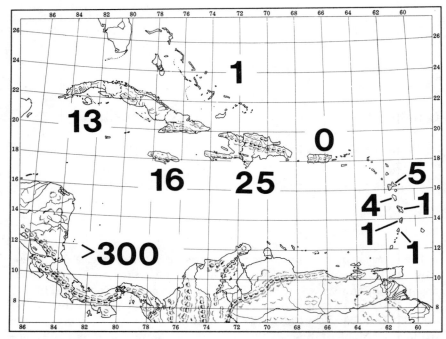

Figure 6-1. Number of *Platynus* beetle species endemic to the Antillean islands. Southern Florida supports no species related to West Indian species. Mexico, Central America, and South America contain over 300 species. (Copyright by American Map Co., Inc.)

(Darlington, 1934, 1935, 1937a, 1937b, 1939, 1953, 1964). Subsequent collecting and a search of all major collections coordinated by Terry L. Erwin of the Department of Entomology at the Smithsonian Institution, Washington, D.C., has resulted in the discovery of 10 more species. Doubtless more species remain undiscovered, and this paper should be considered a progress report on what we know about the fauna.

Methods

In this contribution I present a review of the cladistic relationships of the West Indian *Platynus* species. Analysis of these relationships with regard to the geologic hypotheses proposed for the formation of the Caribbean area suggests tenable and untenable hypotheses for the diversification of this fauna. I adopt the standpoint that finding concordant patterns of relationships within the fauna is evidence of an orderly vicariance of an ancestral biota (Rosen 1978, Nelson and Platnick 1981). Lack of such concordant patterns, and indeed the existence of conflicting patterns of taxon-area relationships, are grounds for hypothesizing dispersal as an agent for colonization of the Antilles. In several instances, I have used geologic data or the overall pattern of diversity of a particular species group to estimate time of origin on the Antilles of that group. I do not view dispersal as a force so pervasive that it makes the search for general patterns a useless pastime. On the contrary, by minimizing ad hoc hypotheses about dispersal we can better consider the circumstances under which dispersal does or does not occur, giving a balanced view to the forces of colonization and fragmentation at work in the Antillean fauna.

Cladistic Analysis

A cladistic analysis groups taxa hierarchically based on shared and derived or synapomorphous characters (Hennig 1966). In this analysis, determination of the derived or primitive states of characters is based on out-group comparison (Watrous and Wheeler 1981), assuming the genus *Agonum* to be the sister group of *Platynus* (Liebherr 1986b). Minimization of the number of character-state changes across the cladogram is used to optimize the cladogram (Sober 1983), through the use of the Phylogenetic Analysis Using Parsimony (PAUP) computer algorithm of Swofford (1984). The computer analysis uses the Min F option for addition of taxa to the network, a root based on a plesiomorphic hypothetical sister group, and local branch swapping to search for the shortest cladogram. The plesiomorphic and apomorphic character states are described in Appendix 6.1. Directions for recreating the taxon × character data matrix are presented in Appendix 6.2. Names of taxa used in this analysis are formalized in Liebherr (1987).

The most parsimonious computer-generated cladogram produced several unacceptable cladistic groupings. In several cases this occurred because characters extensively homoplasious over the entire cladogram grouped taxa optimally within one local area of the cladogram. When this happened, relationships were reassessed based on less homoplasious characters. In another example, a largely plesiomorphic though phenetically homogeneous species group restricted to one island clustered along the bases of various clades emanating from the basal node of the computer-generated cladogram. The monophyly of this group was assumed based on phenetic similarity and on the group's occurrence on a single island. Clearly such a group is not to be preferred when conducting biogeographic analyses.

In order to accommodate the difficulties introduced by adherence to the most parsimonious cladogram, a modified, less parsimonious cladogram is used as the basis for biogeographic analysis of the *Platynus* fauna. The additional assumptions of character-state changes necessary to allow use of this cladogram are included in the results. Use of this modified cladogram recognizes the trade-offs between the use of parsimony as an optimizing criterion, and the necessity to recognize phylogenetic affinity based on less homoplasious characters and, in extreme cases, distributional or phenetic criteria.

Morphological Characters

I coded 66 characters plesiomorphic or apomorphic for the 66 species-level taxa in the analysis. Presence or absence of setae, and their position on the body surface, serve as 19 characters. Setae used as characters are shown in Figure 6-2. If a seta is observed in many platynine carabid beetles, including species of *Agonum*, it is considered primitively present. The loss of a particular seta often assists the definition of natural groups in this fauna, based on correlation of these characters with other characters and the geographic distribution of the lineages.

The microsculpture on the cuticle of the beetles offers a valuable character for species diagnosis, and in some comparisons aids the definition of natural groups. Two basic patterns of microsculpture are shown in Figure 6-3. The terms for carabid body microsculpture were standardized by Carl Lindroth (1974), who used this characteristic to advantage in making easy-to-use keys to the North American ground-beetle fauna (Lindroth 1969 et seq.).

The development of the metathoracic flight wings varies among the West Indian *Platynus*, with many species exhibiting shortened flight wings and others possessing an abbreviated metathorax and only vestigial wings (Darlington 1970). The loss of flight ability is a specialization found predominantly in mountain species (Darlington 1943).

The structure of the tarsi is useful for delimiting several groups. A morphocline in the relative lengths of lobes on the fourth tarsomere is observed in tropical *Platynus* (Fig. 6-4). The outer and inner lobes may be equal and short or long (Fig. 6-4A). Alternatively, the outer lobe may be elongate and the inner lobe short (Fig. 6-4B–D). The tarsi of arboreal species in the *bromeliarum* group of Jamaica are dramatically expanded laterally (Fig. 6-4D, E) and possess many presumably adhesive setae on the ventral surface. These species live in bromeliads, so the tarsal structure is apparently a development useful in arboreal habitats.

Eighteen characters were coded from the male and female reproductive structures. The female reproductive tract, long assumed lacking in characters, exhibits variability in several features. The gonocoxae (Bils 1976), or ovipositor lobes, are divided into a basal and apical segment, with each bearing a number of setae (Fig. 6-5). The structure of the bursa copulatrix, or the receptacle for the male aedeagus, may bear a dorsal sclerite (Fig. 6-5) and may or may not possess a dense inner coating of microtrichia.

In the males, the aedeagus, or intromittent organ, comprises a median lobe and two lateral parameres. Variability in these structures is found mostly in

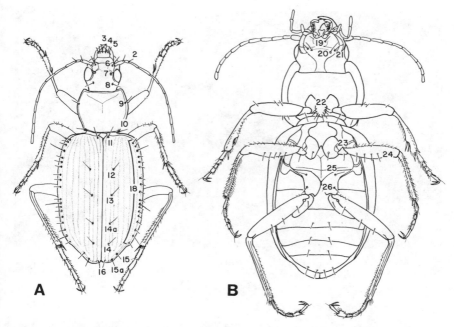

Figure 6-2. Setal numbering scheme used to code variably present setae for cladistic analysis. A. *Platynus punctinotus*, St. Vincent, dorsal view. B. *P. biramosus*, Hispaniola, ventral view.

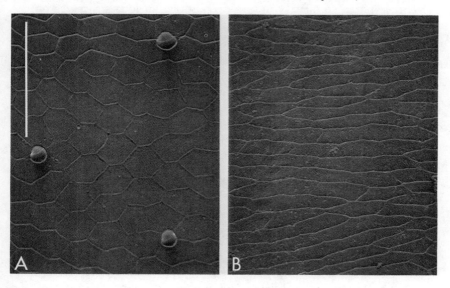

Figure 6-3. Body surface microsculpture on elytra. A. Isodiametric mesh of *Platynus darlingtoni*. B. Transverse mesh of *P. fractilinea* (scale bar, 50 μm).

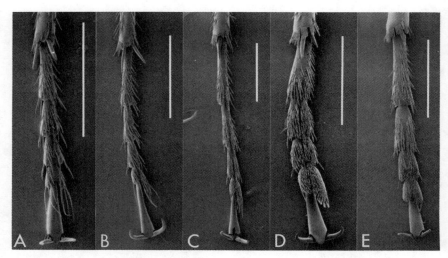

Figure 6-4. Ventral view of hind left tarsus of representative *Platynus*. A. *P. cinchonae*, Jamaica. B. *P. aequinoctialis*, specimen from Jamaica. C. *P. vagepunctatus*, Jamaica. D. *P. bromeliarum*, Jamaica. E. *P. darlingtoni*, Jamaica (scale bar, 1 mm).

the presence of sclerotized teeth on various regions of the internal sac (Fig. 6–6). This sac is normally held folded within the median lobe and is everted into the female bursa copulatrix during copulation. Various structures on the internal sac of tiger beetle males (Carabidae: Cicindelinae) have been implicated in the actual mating process (Freitag et al. 1980). In *Platynus* the only currently demonstrable correspondence between male and female reproductive tracts is a loose correlation between the size of the male internal sac and that of the female bursa.

Results

Overall Pattern of Relationships

The most parsimonious computer-generated cladogram of the West Indian *Platynus* (Fig. 6-7) is highly unresolved at the basal node. This is largely due to the relatively plesiomorphic characteristics of the Cuban species in the *baragua* group. These species include *P. fratrorum, P. baragua, P. pinarensis, P. medius, P. carabiai, P. bucheri,* and four other Cuban species located off the basal node—*P. mediopterus, P. subangustus, P. bruneri,* and *P. turquinensis*. Arising out of this basal rosette of predominantly plesiomorphous species are a number of well-defined monophyletic groups. These include the *bromeliarum* and *cinchonae* groups of Jamaica with 12 and 4 species, respectively. The *memnonius* group (8 species) is found on the Lesser Antilles and

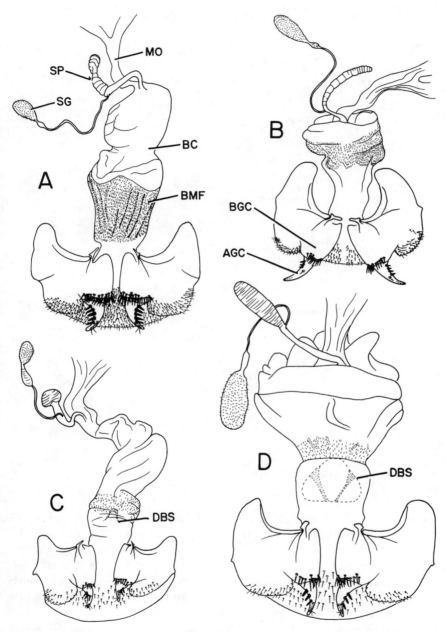

Figure 6-5. Female reproductive tract of *Platynus* in ventral view. A. *P. bruesi*. B. *P. jamaicae*. C. *P. pavens*. D. *P. visitor*. BC, bursa copulatrix; SP, spermatheca; SG, spermathecal gland; MO, median oviduct; BMF, bursal microtrichial field; AGC, apical gonocoxal segment; BGC, basal gonocoxal segment; DBS, dorsal bursal sclerite.

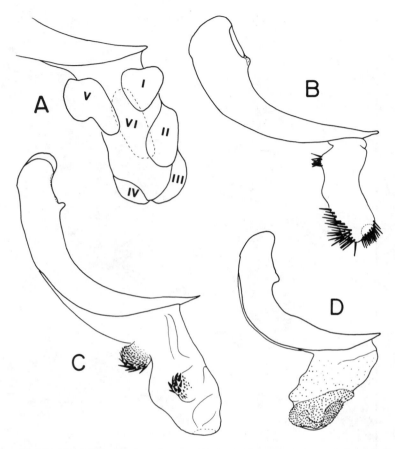

Figure 6-6. Median lobe of male aedeagus with internal sac everted, ventro-lateral view: A. Schematic showing regions of sac coded for presence of sclerotized spines, region VI is dorsal. B. *Platynus pavens*. C. *P. wolla*. D. *P. visitor*.

the mainland, and the *fractilinea* group (6 species) and *laeviceps* group (5 species) on Hispaniola. The *jaegeri* group has 9 species on Hispaniola, and 2 in the mountains of Cuba. The computer-generated cladogram of 66 taxa and 66 single-state characters is 223 steps long, a consistency ratio of .296.

The species of the *baragua* group are restricted to Cuba and are phenetically very homogeneous. Most species can be diagnosed only by subtle differences in pronotal shape, elytral setation, and body microsculpture. If we assume the group is monophyletic (Fig. 6-8), we remove much of the unresolved cluster about the basal node of the most parsimonious cladogram. Such an assumption results in six extra steps over the entire clado-

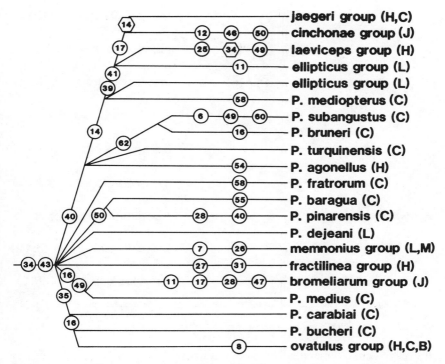

Figure 6-7. Most parsimonious cladogram showing monophyletic species groups, or species where groups are cladistically undefinable. Islands: B = Bahamas; C = Cuba; H = Hispaniola; J = Jamaica; L = Lesser Antilles; M = mainland. Characters are indicated by number, with character state advances in circles, reversals in hexagons.

gram. These character-state changes and the rationale for accepting them are discussed in the *baragua* group presentation below.

Several other taxa are placed in taxonomic groupings of the modified cladogram (Fig. 6-8) not specified in the most parsimonious cladogram (Fig. 6-7). *P. agonellus* is placed in the *ovatulus* group (one extra step), *P. dejeani* in the *memnonius* group (no extra step), and *P. visitor*, which was placed within the *ovatulus* group of Figure 6-7, is considered the most plesiotypic member of the *fractilinea* group (four extra steps). The Lesser Antillean *ellipticus* group is shown as a single lineage in Figure 6-8, even though no morphological characters support the monophyly of the group (two extra steps). The rationale for each of these decisions is presented with each species group below.

Overall, the modified cladogram is 13 steps longer than the most parsimonious cladogram and has a consistency ratio of .280. Thus we have increased the length of the cladogram 5.8% in an attempt to better represent

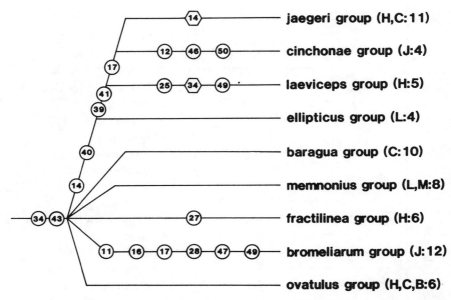

Figure 6-8. Simplified cladogram showing relationships and defining character-state transformations for species groups. Such a cladogram assumes species groups are monophyletic, but in several groups no such synapomorphies are known.

patterns of character change and to maintain geographically coherent species groups in a taxon replete with insular species.

The structure of the modified cladogram (Fig. 6-8) principally involves the restriction of species groups to single islands, or island groups. The only resolved species group relationships include the *ellipticus, laeviceps, cinchonae,* and *jaegeri* groups. Advances in characters 14 and 40 (pronotal hind seta placement and flight-wing reduction) distinguishes the *ellipticus* group and its sister clade from the basal node. The *laeviceps-cinchonae-jaegeri* clade is distinguished by complete reduction of flight wings to vestiges (character 41), and reduction of the metathorax (character 39). Such reduction in the metathoracic flight apparatus occurs repeatedly throughout the Platynini, as well as in other carabid groups. The *cinchonae-jaegeri* clade is united by the occurrence of broadly reflexed pronotal margins (character 17). If we are to accept the monophyly of this set of species groups, we must bear in mind that few characters support such an assessment.

Monophyly of West Indian *Platynus*

Without doubt, the West Indian *Platynus* fauna is not monophyletic. Several components of the fauna have clear mainland relatives. The *ovatulus* group is distributed throughout Central America and Mexico, and the species on the Antilles represent an extension of that group. Similarly, the *memnonius* group contains *P. aequinoctialis*, a species found on the mainland, Cuba, Hispaniola, and Jamaica. The relationships of other distinctive West Indian groups to mainland groups are not known.

That we can only do cladistic analysis on a monophyletic group is a misconception often used to delay the introduction of such analysis in inadequately known groups. Cladistic analysis groups taxa in a relative manner based on shared-derived states of characters. If character states are properly polarized, the relative relationships of taxa in an analysis will not be disturbed by the addition of new taxa. The addition of new characters will change the relative relationships only if they can outweigh the groupings based on the original characters.

Using the cladogram in Figure 6-8 for testing biogeographic hypotheses, while knowing the group is not monophyletic, simply means we must restrict our discussion to groups shown to be monophyletic on the cladogram, that is, those species groups defined by synapomorphies. Even if these limits are placed on the interpretation of the results, a much clearer picture of the composition of the fauna can be obtained using these methods. Below I describe the faunas of the various islands, along with the likely relationships.

The Hispaniolan Fauna

Platynus is represented on Hispaniola by four species groups (Figs. 6-7, 6-8), which have resulted from three ancestors. The *jaegeri* group comprises heavy-bodied beetles with wide prothoracic margins and a reduced metathorax (Figs. 6-9, 6-10). The *laeviceps* group is closely related (Figs. 6-8, 6-10) and composed of less robustly built species. *P. constricticeps* of the *jaegeri* group is phenetically similar to species of the *laeviceps* group, and were it not for the presence of the *cinchonae* group of Jamaica, the *laeviceps* and *jaegeri* groups would be considered one species group (Figs. 6-8, 6-10).

The species of the *jaegeri* and *laeviceps* groups possess very restricted distributions. All species are restricted to single mountain systems. In the *laeviceps* group, species are restricted to either the mountains of the Presqu'Ile du Sud of Haiti (Fig. 6-10, SH) or the Cordillera Central-Massif du Nord (Fig. 6-10, CH). Similar restricted distributions are exhibited by species of the *jaegeri* group. *P. biramosus*, *P. christophe*, *P. cychrinus*, and *P. calathinus* occur in the Cordillera Central-Massif du Nord system; *P. transcibao* is found in the northern-most range, the Cordillera Septentrional; and *P. tipoto*, *P. subcordens*, *P. jaegeri*, and *P. constricticeps* are restricted to the Presqu'Ile du Sud of Haiti. The two Cuban species of the *jaegeri* group also are of restricted distribution: *P. acuniai* is known only from Pico Turquino of Oriente Province, and *P. cubensis* from the Trinidad Mountains of Las Villas Province.

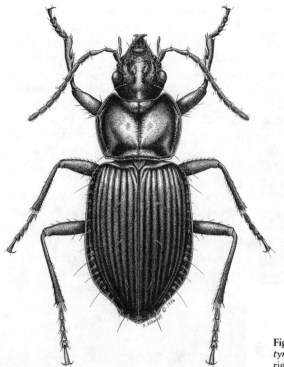

Figure 6-9. Dorsal habitus of *Platynus jaegeri*, Hispaniola. (Copyright © 1986 by K. Schmidt.)

The *fractilinea* species group (Fig. 6-11) is a generalized group of species; four of the six possess fully developed flight wings. They are little differentiated from the *Platynus* ground plan, with the exception of the unique reduction of the elytral striae to dashlike grooves.

In the computer PAUP analysis (Fig. 6-7), *P. visitor* was placed in the *ovatulus* group. The most complex and least homoplasious character supporting such a hypothesis is the presence of a bursal sclerite (character 54, Fig. 6-5C, D). Positionally, the various examples of this character state are homologous, but the structures shown in Figure 6-5C and D are quite different. In addition, the apparently independent origin of bursal sclerites in *P. marcus* and *P. wolla* of the *laeviceps* group shows that this character is prone to homoplasy. *P. visitor* is much larger than species of the *ovatulus* group (9.0–10.5 mm versus 6.6–8.5 mm), and it is much more robustly built. Body size and robustness were not scored cladistically because of the very small sample sizes in some species for the former and the difficulty in making an overall qualitative coding of body build for the latter. Nonetheless, based on these attributes, I believe *P. visitor* is misplaced by the com-

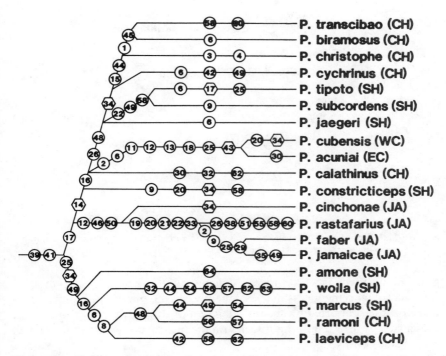

Figure 6-10. Cladistic relationships of species in *Platynus laeviceps, cinchonae,* and *jaegeri* species groups. The *laeviceps* group comprises bottom 5 species; *cinchonae* group the 4 Jamaican species; *jaegeri* group the top 11 species. Characters labeled by number; character state advances in circles, reversals in hexagons. Distribution of each species shown in parentheses: CH = central or northern Hispaniola; SH = southern Hispaniola; WC = western Cuba; EC = eastern Cuba; JA = Jamaica.

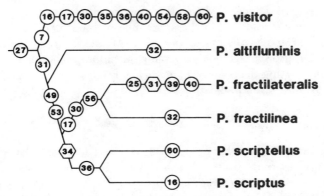

Figure 6-11. Cladogram of *Platynus fractilinea* species group; character state advances in circles, reversals in hexagons.

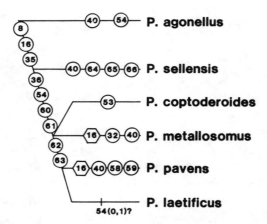

Figure 6-12. Cladogram of West Indian species of *Platynus ovatulus* species group; character state advances in circles, reversals in hexagons. State of character 54 unknown for *P. laetificus*.

puter analysis when it is considered a member of the *ovatulus* group. A more satisfactory placement is as the basal member of the *fractilinea* group. Such a placement is supported by shared reduction of the elytral striae and by the presence of short versus long spicules on the male aedeagal internal sac (Fig. 6-6C versus 6-6B).

The *ovatulus* group is represented by 6 species in the West Indies (Fig. 6-12). This group is widespread in the mainland Neotropics, with at least 15 mostly flighted species found from southeastern Arizona to South America. Five of the West Indian species are restricted to Hispaniola, whereas *P. coptoderoides* occurs on Cuba and the Bahamas. Bahamian records for *P. coptoderoides* have included capture at light.

P. agonellus is grouped with several Cuban species (Fig. 6-7) based on the synapomorphy of reduced flight-wing length. But this species is smaller than the Cuban species and within the size range of the other 5 species in the *ovatulus* group. It also possesses a dorsal bursal sclerite seen in other species of the *ovatulus* group (Fig. 6-5). Assuming the minimum number of dispersal events in populating the islands, *P. agonellus* is considered a member of the *ovatulus* group (Figs. 6-8, 6-12).

The Jamaican Fauna

The Jamaican *Platynus* have resulted from two radiations. The *cinchonae* group comprises four species, which possess vestigial wings and reduced metathoraxes, and are the sister group to the *jaegeri* species group of Hispaniola (Figs. 6-8, 6-10). These groups are the sister to the *laeviceps* group of Hispaniola, implying that the ancestor of the *cinchonae* group was separated from a Hispaniolan lineage and subsequently radiated on Jamaica.

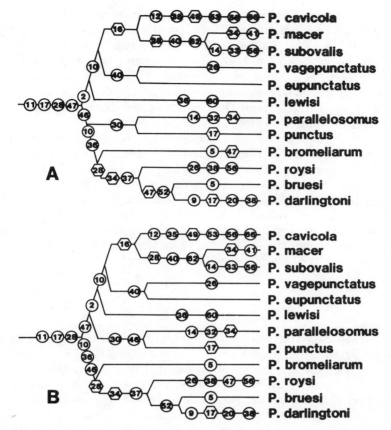

Figure 6-13. Two equally parsimonious cladograms for the *Platynus bromeliarum* species group; character state advances in circles, reversals in hexagons. A. Preferred cladogram based on evolution of tarsi. B. Less preferred but equally parsimonious cladogram.

The other group inhabiting Jamaica is the *bromeliarum* group of 12 species. This stock exhibits a classical island radiation, with a variety of habitats used by the various species. Two hypotheses of relationships are equally parsimonious (Fig. 6-13), each requiring 48 character-state changes. The difference between these two cladograms lies in the placement of the root or basal node. In one cladogram (Fig. 6-13A) the root is placed between *P. lewisi* and the stem leading to *P. punctus* and *P. parallelosomus*. Based on this cladogram, the tarsi have evolved in a transformation series of B (= C) → D → E (Fig. 6-4). Based on the modified cladogram (Fig. 6-8), the common ancestor of the *bromeliarum* group had tarsi with the fourth tarsomere bearing an outer lobe more than twice as long as the inner lobe (character 47). During the radiation on Jamaica, one character-state advance involved tarsi becoming broader in one clade (*P. punctus* and relatives, Fig. 6-4D, E;

Fig. 6-13A). In several species of this line (*P. bromeliarum*, *P. bruesi*, and *P. darlingtoni*), the outer lobe of the fourth tarsomere is shorter than twice the length of the inner lobe (Fig. 6-4E). This is considered a reversal in character state, back to that coded primitive for the relative lengths of the fourth tarsomere lobes (as in Fig. 6-4A). All species in the *P. bromeliarum* clade possessing broadened tarsi on the meso- and metalegs exhibit an abundance of setae on the ventral surface. These would appear to be an adaptation to arboreal life providing sure footing, as all species with broad tarsi are usually taken inside bromeliad epiphytes. The species with broad tarsi are always fully winged, necessary for a largely arboreal existence. Other characteristics of these species include dorso-ventrally flattened bodies, elongate heads, metallic body coloration, and spines on the tips of the elytra.

A second hypothesis of *bromeliarum* group relationships requires either that broadened tarsi evolved in the common ancestor of the *bromeliarum* group and were later lost in *P. lewisi* and its more derived relatives, or that broadened tarsi (character 46) evolved twice (Fig. 6-13B). The transformation series for the first alternative would be expressed as

$$B \rightarrow D \begin{array}{c} \nearrow C \\ \searrow E \end{array}$$

(Fig. 6-4). Broadened tarsi on the hindlegs are observed in few groups within the genus *Platynus*. To hypothesize origin and then loss, or two origins of such a novel structure, is less parsimonious than assuming it arose once.

Besides the primarily arboreal species, several other types of specialization have occurred in this group. Four species exhibit shortened flight wings (character 40) and cannot fly (*P. vagepunctatus*, *P. eupunctatus*, *P. macer*, and *P. subovalis*). Judging from the presence of fully developed flight wings in *P. cavicola*, brachyptery has evolved twice. Even though these species cannot fly, several of them have been collected in bromeliads. Of the six species in the *P. lewisi* clade, *P. lewisi*, *P. macer*, and *P. vagepunctatus* have been recorded in bromeliads. This suggests epiphyte habitation is a primitive characteristic for this group, with brachyptery playing a small role in restricting habitat utilization by these beetles.

Interestingly, the single West Indian *Platynus* taken in caves possesses well-developed eyes and is fully winged. *P. cavicola* has been taken to date in two caves in the Cockpit country of northern Jamaica. Whether this species is a troglophile found in caves due to the nature of the terrain, or is a troglobite restricted to caves, is not known.

The Cuban Fauna

The generally plesiomorphic and phenetically homogeneous *baragua* group comprises 10 of the 14 Cuban *Platynus*. The cladistic relationships of

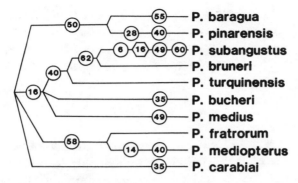

Figure 6-14. Cladogram of *Platynus baragua* species group; character state advances in circles, reversals in hexagons. Basal node is equivalent to basal node of cladogram in Figure 6-7.

these species relative to all West Indian *Platynus* are shown on the most parsimonious cladogram (Fig. 6-7). If we assume this group has radiated in isolation on Cuba, their relative relationships are as shown in Figure 6-14. In either case, the relationships within the group are highly unresolved.

One sister-species pair is supported by a derivation in the gonocoxal setation (character 50): *P. baragua* and *P. pinarensis*. This may be interpreted as a vicariant sister pair, as *P. pinarensis* is restricted to the mountains of Pinar del Rio and *P. baragua* inhabits the Trinidad Mountains in Central Cuba.

The other species in this group are known mostly from various peaks in the Sierra Maestra of eastern Cuba. The species generally exhibit limited distributions, with each mountain system possessing one or more riparian plus more forest-adapted geophilic species (Darlington 1970). The occurrence of stream-side and forest-adapted species within mountain systems suggests ecological differentiation as a major factor in this radiation.

The two species of the *jaegeri* group, *P. cubensis* and *P. acuniai*, constitute the second clade of Cuban *Platynus*. These are sister species, sharing a number of synapomorphies (Fig. 6-10). Their relationship is no doubt due to isolation in mountain habitats subsequent to a more cosmopolitan ancestral distribution.

P. coptoderoides respresents the *ovatulus* group in eastern Cuba and throughout much of the Bahamas.

The Lesser Antillean Fauna

The *Platynus* species of the Lesser Antilles can be placed in two species groups based on the cladistic analysis (Figs. 6-7, 6-8): the *ellipticus* group and the *memnonius* group. The *ellipticus* group is comprised of four species, which share a derived placement of the hind-pronotal seta (character 14) and

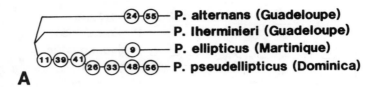

A

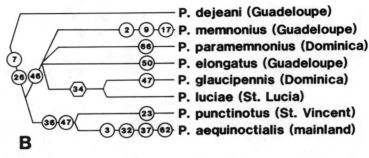

B

Figure 6-15. Cladograms of two lineages inhabiting the Lesser Antilles; character state advances in circles, reversals in hexagons. Distributions of species shown parenthetically. A. *Platynus ellipticus* species group. B. *P. memnonius* species group.

hindwing reduction (character 40) with several Greater Antillean groups. Within the *ellipticus* group, complete reduction of flight wings is observed in the species from Dominica and Martinique (characters 39, 41; Fig. 6-15A), whereas the two Guadeloupan species possess partially developed brachypterous wings. Such a finding suggests progressive derivation from north to south, that is, Hennig's "Progression Rule" (1966). Progressive loss of vagility in a clade is not the type of character usually used in the progression rule, as the argument can be made that longer-term residents of an island should exhibit greater reduction in dispersal powers. Moreover, the *ellipticus* group cannot be monophyletically defined based on structural characters (Fig. 6-7), nor can *P. lherminieri* be distinguished from the common ancestor of the group, which is also the basal node of the *laeviceps-cinchonae-jaegeri* clade (Figs. 6-8, 6-15). At the level of resolution of the characters, all we can conclude is this group of species shows affinities to Greater Antillean taxa.

The *memnonius* group comprises eight species, one of which can be found on the mainland (Fig. 6-15B). *P. dejeani* is included in this group for convenience, even though its overall plesiomorphic nature does not support such a conclusion on cladistic grounds. In this group, species from adjacent islands tend to be related. The closer relationship of *P. punctinotus* to the mainland *P. aequinoctialis* compared to its relationship to species found on the more northern islands suggests at least two invasions of these oceanic islands from South America.

Table 6-1. Diversity of species and species groups on the Antillean islands, and the percentage of brachypterous species and lineages in each island system

Island	No. endemic species	Total no. species	No. of lineages*	No. spp. in each lineage*	No. spp. w/wing reduction (%)*	No. lineages w/wing reduction (%)
Jamaica	16	17	2	12, 4	8 (50)	2 (100)
Cuba	12	14	2	10, 2	7 (58)	2 (100)
Hispaniola	25	26	4	9, 6, 5, 5	19 (76)	4 (100)
Bahamas	0	1	0	—	0 (0)	0 (0)
Lesser Antilles	11	11	2	7, 4	4 (36)	1 (50)

*Based on endemic species.

Diversity and Brachyptery

With four lineages (three if the *laeviceps* and *jaegeri* groups are united) and 25 endemic species, Hispaniola far and away contains the most diverse *Platynus* fauna in the West Indies (Table 6-1). The distribution of species among the lineages is more harmonic than observed on the other islands, with 9, 6, 5, and 5 species (or 14, 6, and 5). The percentage of species exhibiting some form of flight-wing reduction is highest on Hispaniola (Table 6-1), with 19 of the 25 species exhibiting reduced vagility. As reduced dispersal capacity can be related to increased habitat stability (Southwood 1962), these findings suggest Hispaniola has been relatively more stable climatically than any of the other Greater Antilles. The large area and greater topographic range of this island would allow the resident fauna opportunity to track specific habitats during periods of climatic change. The ability to respond to climatic change through altitudinal shifts would reduce extinction and enhance speciation, as species ranges would be fragmented in upland areas during drier climatic regimes.

Half or slightly more of the endemic species on Jamaica and Cuba are brachypterous. On the three Greater Antillean islands, brachyptery occurs in all lineages containing endemic species.

The Lesser Antillean *Platynus* fauna is comprised of only 11 species, with brachyptery restricted to the *ellipticus* group of four species. The other group is comprised exclusively of macropterous species. Brachypterous species are limited to the northern islands of Guadeloupe, Martinique, and Dominica.

Discussion

Derivation of the *Platynus* fauna

The number of phyletic stocks that established the Antillean *Platynus* fauna will remain uncertain until the mainland fauna is analyzed cladistic-

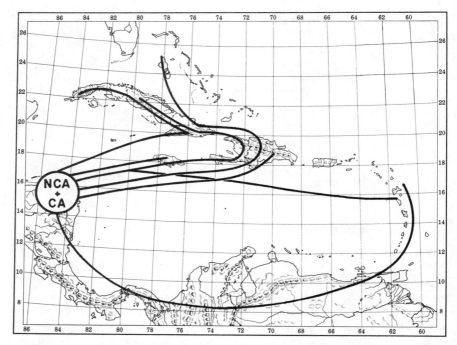

Figure 6-16. Individual tracks for species groups of West Indian *Platynus*. Northern Central America (NCA) and Central America (CA) considered the likely areas containing groups related to Antillean groups. (Copyright by American Map Co., Inc.)

ally. The number of lineages containing endemic species on each island system (Table 6-1) can be obtained from the present analysis. Jamaica, Cuba, and the Lesser Antilles possess two such lineages each, whereas Hispaniola has four lineages, established from three progenitors.

The ranges of distribution of each species group, also called individual tracks (Rosen 1975), suggests that a mixture of overwater dispersal and vicariance has led to the present fauna (Fig. 6-16). Northern Central America, including Mexico, and Central America are considered the likely areas containing lineages most closely related to Greater Antillean *Platynus*. As we learn more about the mainland fauna, such assumptions may require reconsideration.

The Hispaniolan fauna contains the *ovatulus* group, which has also radiated on the mainland, and which contains one species that has colonized the Bahamas from Cuba. A cladistic analysis of all species in this group will determine whether the Antillean species form a monophyletic group. If *P. agonellus* remains at the cladistic base of an Antillean species clade when the entire *ovatulus* group is analyzed (Fig. 6-12), a vicariant origin of this group will be supported. This conclusion results because *P. agonellus* is relatively plesiomorphic within the *ovatulus* group, and its presence at the base of a West Indian clade would suggest early divergence of that clade, consistent with Eocene vicariance (Rosen 1985). An overwater origin of the group will be supported if analysis of all *ovatulus* group species shows the more recently derived species, such as *P. metallosomus*, *P. pavens*, and *P. laetificus*, to be members of a separate lineage than that containing *P. agonellus*. Such a finding would suggest that the former three species are members of a terminal group of relatively recent origin, making the Antillean representatives a likely result of overwater colonization.

The other three Hispaniolan groups, the *fractilinea*, *laeviceps*, and *jaegeri* groups, can be derived from at most two ancestors. Based on this analysis (Fig. 6-7), an independent origin of the *fractilinea* group is assumed, but study of mainland groups is necessary to determine this.

The *laeviceps* and *jaegeri* groups form two parts of what I will hereafter call the wingless clade, with the Jamaican *cinchonae* group forming the third part. If vicariance has acted on an ancestral *Platynus* fauna, this group should be a prime example because the member species exhibit reduced vagility and extremely parochial distributions. Rosen (1985) presented a tectonic model of Caribbean biogeography containing four stages, starting in the early Eocene, about 50 Ma (Fig. 6-17, stages 1–4). This model is based on the most recent mobilist geologic hypotheses for the Caribbean, such as Sykes et al. (1982) and Wadge and Burke (1983). If we reduce Figure 6-10 to a taxon-area cladogram and fit it to Rosen's model, we obtain a scenario combining vicariance and dispersal (Fig. 6-17, top). The initial vicariance in this clade is dated early Eocene, when the ancestors of the *laeviceps* and *cinchonae* groups colonized southern Hispaniola and Jamaica from northern Central America (stage 1). Subsequently Jamaica and southern Hispaniola separated from northern Central America (stage 2) as the Caribbean plate moved eastward relative to North and South America. In order for the *jaegeri* group to colonize southern and central Hispaniola, two dispersal events must be hypothesized if this lineage originated after Eocene movement of the Caribbean plate (stage 1 to 2). Subsequently, the central Hispaniola–eastern Cuba island vicariated due to strike-slip faulting along the northern edge of the Caribbean plate (stage 3), isolating the ancestor of the present two Cuban species of the *jaegeri* group. In the late Pliocene, southern Hispaniola sutured onto central Hispaniola, allowing range expansion (stage 4), with Pleistocene climatic oscillations producing the present-day species.

This tectonic model appears to account for diversification within this lineage by a combination of vicariance and linear island hopping during the Eocene (stage 2). It also suggests the 20 species of the wingless clade have been the result of 50 Ma of evolution. However, there is one geologic constraint that calls into question the credibility of this model. Even though

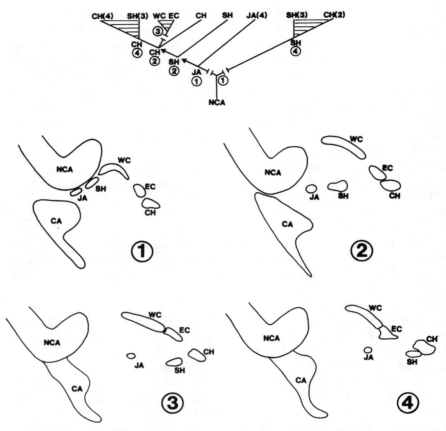

Figure 6-17. Reduced taxon-area cladogram of the wingless clade (*Platynus laeviceps-cinchonae-jaegeri* groups of Fig. 6-10) interpreted using Rosen's (1985) tectonic model of Caribbean biogeography. Numbers of species represented on reduced cladogram shown parenthetically if greater than one. Taxon-area cladogram on top utilizes vicariance (⊢⊣⊢), dispersal (→), and range expansion (⟱). Nodes of area cladogram are numbered, corresponding to configurations of land masses in stages 1–4 of Rosen's hypothesis. Stages shown at bottom are (1) early Eocene, approximately 50 Ma; (2) late Eocene, 35 Ma; (3) late Miocene, 12–20 Ma; (4) late Pliocene to present, 0–5 Ma. Land-mass abbreviations as in Figure 6-10, with NCA = northern Central America; CA = Central America.

Jamaica contains some very old rocks, dated upwards of 100 Ma (Donnelly 1985), and was above sea level in the early Eocene, extensive subsidence from late Eocene through mid-Miocene (Robinson et al. 1970, Arden 1975, Buskirk 1985) resulted in deposition of limestone that is characteristic of complete insular submergence at that time (Steineck 1974). If this interpretation of the data is accepted, the diversification of this group must start at stages 2–3 (Fig. 6-17), with overwater colonization of Jamaica, southern, and then central Hispaniola followed by vicariance of central Hispaniola–eastern Cuba, the suturing of eastern and western Cuba, and finally the suturing of southern and central Hispaniola. If the Nicaraguan rise was emergent in the Miocene, such overwater colonization could have occurred over a much shorter distance than would be required today (see Donnelly, chap. 2).

The origin of the *laeviceps* group via Eocene vicariance is not countermanded by geologic data. The southern peninsula of Hispaniola, the Presqu'Ile du Sud-Baoruco block, is formed from uplifted sea floor (Lewis 1980), with terrigenous clastics indicating existence of terrestrial habitats for an indeterminate period before, and at least since mid-Miocene (Maurrasse et al. 1980). However, corroboration of Eocene vicariance by what is essentially negative evidence is hardly a preferred state of affairs.

Two other species groups possess tracks composed of a single island and the mainland. For the *baragua* group of Cuba, the overall phyletic similarity of the species and the restricted species' distributions are the bases for regarding this group as a recent radiation. Such a time of origin would be most logically based on overwater colonization, probably from Central America (Fig. 6-16). Placing this group in a cladistic analysis containing mainland groups would substantiate this conclusion.

The *bromeliarum* group of Jamaica is well defined cladistically (Figs. 6-7, 6-8), but its relationships remain obscure (Fig. 6-16). The most parsimonious cladogram places the group as the sister to *Platynus medius* of the *baragua* group (Fig. 6-7). The characters supporting such placement (pronotal shape, 16; number of gonocoxal setae, 49) are extensively homoplasious, changing state 10 and 9 times, respectively, over the entire cladogram. There is little reason to suspect *P. medius* to be the sister group. A search of mainland *Platynus* lineages may bring a likely sister group to light, especially because the *bromeliarum* group is characterized by a number of distinctive apomorphies. Based on sedimentary data (Steineck 1974) indicating that Jamaica has been permanently emergent only since the mid-Miocene, the *bromeliarum* group is also considered the result of overwater dispersal. As mentioned above, such overwater colonization may have been of a short distance, due to an emergent Nicaraguan rise (see Donnelly, chap. 2).

The Lesser Antilles host two species groups. One has Greater Antillean affinities and is found on the northern islands, which contain rocks of the first cycle of volcanism in the Lesser Antillean island arc, and the second is found throughout the island chain with affinities to mainland South America (Fig. 6-16). The northern islands of Guadeloupe Basse Terre—La Desirade, parts of Martinique and St. Lucia, and probably part of Dominica—have basement rocks dated pre-Miocene (Fink 1970). Thus, existence of northern islands colonized from the Greater Antilles could have occurred before colonization of the island chain from the south by the *memnonius* group.

Why Are There No *Platynus* on Puerto Rico?

Despite numerous collectors on the island, extensive fieldwork by P. J. Darlington, Jr., and J. Ramos, and a field trip by the author specifically to look for them, no *Platynus* beetles have been taken on Puerto Rico. Although Puerto Rico has forest habitats and mountains of a stature proper for beetles of this genus, the group appears to be absent.

If we observe that the West Indian *Platynus* fauna is the result of relatively few founders (Table 6-1), the apparent absence of *Platynus* from Puerto Rico becomes understandable. The island is geographically isolated at the eastern end of the Greater Antilles, which, combined with its small size, make Puerto Rico a comparatively poor candidate for receiving overwater immigrants. Because successful colonization events are rare for *Platynus*, it is not surprising that this island has not received such a colonist.

The theory that *Platynus* species colonized Puerto Rico and subsequently became extinct can be rejected because of two paleoendemic carabid groups that are present in forest habitats on the island. *Barylaus estriatus* is a flightless, highly derived member of the carabid tribe Pterostichini, subtribe Caelostomi (Liebherr 1986a). The species occurs in mesic to moist habitats from 900 m elevation upward in the Cordillera Central. *Barylaus* is most closely related to African taxa of the Caelostomi, indicating that these beetles have been on Puerto Rico since the Cretaceous breakup of Africa and South America (Anderson and Schmidt 1983, Donnelly 1985). Three flightless endemic species of the genus *Antilliscaris*, of the carabid tribe Scaritini, are also found in montane forest habitats on Puerto Rico. *Antilliscaris* is also most closely related to African taxa and is probably the result of Cretaceous isolation (see Nichols, chap. 5). The fact that flightless paleoendemics have been able to survive in forests typical of *Platynus* habitation on other islands suggests that *Platynus* never got to Puerto Rico.

Both *Antilliscaris* and *Barylaus* are also found on Hispaniola. Sykes et al. (1982) suggested that Puerto Rico, Hispaniola, and eastern Cuba would have been adjacent in the Eocene before extensive movement occurred on the Cayman trench. The distributions of these two genera are likely due to Hispaniola-Puerto Rico vicariance, set at 35–38 Ma (Buskirk 1985). Such a date sets the earliest time that *Platynus* could reach Hispaniola yet be denied entry via land to Puerto Rico. This result is compatible with the wingless clade scenario derived from Rosen's (1985) vicariance model, discussed above.

Comparisons with Other Taxa

There are numerous examples of taxa with Hispaniola as the center of Antillean diversity. Schwartz (1978) presented data for amphibians and reptiles, showing that Hispaniola supports the greatest Antillean diversity of *Eleutherodactylus* frogs (Leptodactylidae), *Sphaerodactylus* gekkos (Gekkonidae), and a colubrid snake fauna comprised of four endemic genera, and overall 8 of the 12 Antillean genera of Colubridae.

For butterflies, both Scott (1972) and Brown (1978) compared and contrasted the endemism of each island, and although their totals of endemic species differ because of differing taxonomic interpretations, both rank Hispaniola as the most diverse of the Greater Antilles. Hispaniola is the center of diversity for one of the few endemic Antillean genera of butterflies—*Calisto* of the Satyridae, or wood satyr family. Munroe's (1950) revision of *Calisto* recognized 20 named species, 15 of which occur on Hispaniola. As in *Platynus*, *Calisto* is largely restricted to forested montane habitats, and species distributions are often very restricted. In several cases, species or subspecies are restricted to the Presqu'Ile du Sud-Baoruco ranges on the southern part of the island, with sister species found in the Cordillera Central. Munroe concluded that speciation in this group is of the vicariant type, with changing climatic regimes alternately permitting dispersal, and then facilitating allopatric speciation.

The apparent overwater colonization of various islands by *Platynus* beetles is also observed for other insect and vertebrate taxa, for example, the sweat bees, or Halictidae (see Eickwort, chap. 10), beetles of the family Carabidae, tribe Scaritini (see Nichols, chap. 5), and for vertebrates such as bats (Baker and Genoways 1978), and snakes of the family Colubridae (Cadle 1985).

Certain patterns of relationship can be observed in various taxa that have apparently colonized the Antilles via dispersal. The most prevalent of such dispersal patterns involves colonization of the Lesser Antilles from South America. Examples of Lesser Antillean taxa derived from South America can be found in the Odonata (Flint 1978), *Hydraena* beetles of the family Hydraenidae (Perkins 1980), butterflies (Scott 1972, Brown 1978), among many others. Relatively few examples of sister-group relationships between northern Lesser Antillean and Greater Antillean taxa are known. The butterfly family Lycaenidae has 9 of the 12 Lesser Antillean species derived from the Greater Antillean fauna (Clench 1963). Among reptiles, the *bimaculatus* subseries of *Anolis* (Iguanidae), found on Dominica, Guadeloupe, and the Leeward Islands, is the sister group to the *cristatellus* subseries found on the Virgin Islands, Puerto Rico, Hispaniola, the Bahamas, and Florida (Gorman et al. 1980).

In addition to the *bromeliarum* group, Jamaica hosts several other bromeliad-inhabiting taxa, including hylid frogs (Trueb and Tyler 1974), grapsid crabs of the genus *Metapaulias* (Chace and Hobbs 1969), and the immature stages of *Erythrodiplax* dragonflies (Libellulidae) and *Diceratobasis* damselflies (Coenagrionidae) (T. W. Donnelly, pers. comm.). All four Jamaican

Hyla frogs live and breed in bromeliads, even though these species represent three separate lineages within *Hyla*. *Metapaulias depressus* is an extremely flat land crab that is restricted to bromeliads in montane habitats. Such prevalence of bromeliad habitation suggests Jamaica underwent severe dry periods since these animals originated on the island. The karst topography of the north-central portion of the island includes habitats in which soil moisture is limited under present-day conditions. But the bromeliad specialists are not now restricted to habitats in the karst areas, suggesting that a historical condition that pervaded the island favored the evolution of epiphyte habitation.

There are indications that epiphyte habitation occurs in other *Platynus* species groups on other islands. *P. glaucipennis* of the *memnonius* group has been taken in the base of *Euterpe globosa* (Palmaceae) fronds on Dominica. Five of the species in the *memnonius* group possess expanded tarsi on the hindlegs (character 46, Fig. 6-15B) that are also present in some of the *bromeliarum* group species. Further research on the habits of *Platynus* species should also tell us more of their biotic history.

Summary

The 66 species of West Indian *Platynus* are arrayed in a limited number of species groups, and most of the species are single-island endemics. The Hispaniolan fauna, with 4 species groups and 25 species is the most diverse on the Antilles. Other islands have fewer groups comprised of endemic species; Cuba and Jamaica have two groups each. The Lesser Antillean chain is populated by two groups, one apparently colonizing the chain from the north via the Greater Antilles, and one colonizing from South America. Puerto Rico lacks *Platynus* beetles. The occurrence of the paleoendemics *Barylaus estriatus* and *Antilliscaris* spp. on Puerto Rico in habitats similar to those occupied by *Platynus* on other islands suggests that extinction has not caused the apparent lack of Puerto Rican *Platynus*. It is more likely that Puerto Rico was isolated before the radiation of *Platynus*, and the island has not received successful overwater colonists since.

Cladistic analysis of the West Indian *Platynus* does not produce concordant or even resolved patterns of relationships among the majority of the nine species groups on the islands (Fig. 6-8). The relationships of the three terminal taxa of the single resolved clade, the *laeviceps-cinchonae-jaegeri* groups, are analyzed in light of Rosen's (1985) tectonic model of Caribbean biogeography. Accepting Rosen's model, diversification of this clade requires three vicariant events, two dispersal events, and three bouts of range expansion. Introducing geologic data supporting a late Eocene to mid-Miocene submergence of Jamaica requires an additional overwater colonization at the expense of one vicariant event. Rosen's model adequately explains taxon relationships among species on Cuba and southern and central Hispaniola, implying that islandic vicariance and hybridization have been at work along the northern edge of the Caribbean plate.

Appendix 6.1: Polarized Character States Used in Cladistic Analysis

Following are descriptions of the plesiomorphic (0) and apomorphic (1) character states for the 66 characters used in this analysis. All determinations of polarity have been based on out-group comparison.

The Head

Character 1: setae 3–5 (Fig. 6-2A) may be of nearly equal length (0), or the inner pair of setae may be reduced (1). *Character 2*: seta 7 (Fig. 6-2A) may be present (0) or absent (1). *Character 3*: seta 19 (Fig. 6-2B) may be single (0), or there may be more than one seta on the mentum (1). *Character 4*: seta 20 (Fig. 6-2B) may be single (0), or there may be many medial setae on the submentum (1). *Character 5*: seta 21 (Fig. 6-2B) may be present (0) or absent (1). *Character 6*: eyes may be fully developed (0) or reduced (1). *Character 7*: lateral depressions on the frons may be shallow (0) or well excavated (1) . *Character 8*: the cervical region, or neck, may be constricted (0) or not constricted (1). *Character 9*: the median mentum tooth may be blunt or unidentate (0), or bidentate (1). *Character 10*: the lacinial galea may bear a thick brush of fine setae (0) or thick peglike setae interspersed between fine setae (1).

The Prothorax

Character 11: seta 9 (Fig. 6-2A) may be present (0) or absent (1). *Character 12*: seta 10 (Fig. 6-2A) may be present (0) or absent (1). *Character 13*: the basal and lateral beads of the pronotum may be continuous (0) or not continuous (1). *Character 14*: seta 10 may be situated at the hind pronotal angle (0), or well before the hind angle (1). *Character 15*: prosternal setae (seta 22, Fig. 6-2B) may be absent (0) or present (1). *Character 16*: the pronotum may be cordate, with posteriorly sinuate lateral margins (0), or quadrate (1). *Character 17*: pronotal lateral margins may be thin, unexpanded (0), or widely reflexed, expanded laterally (1). *Character 18*: the laterobasal pronotal impressions may be present (0) or absent (1).

The Elytra

Character 19: seta 11 (Fig. 6-2A) may be present (0) or absent (1). *Character 20*: seta 12 may be present (0) or absent (1). *Character 21*: seta 13 may be present (0) or absent (1). *Character 22*: seta 14 may be present (0) or absent (1). *Character 23*: seta 14a may be absent (0) or present (1). *Character 24*: setae may be absent on elytral intervals 5 and 7 (0) or present (1). *Character 25*: seta 16 may be present (0) or absent (1). *Character 26*: there may be 14–17 lateral elytral setae (0), or 18 or more (state 1, setae 18, Fig. 6-2A). *Character 27*: elytral striae may be complete to the elytral apex (0) or reduced, incomplete at tip (1). *Character 28*: striae may be smooth (0) or punctate at base (1). *Character 29*: striae may be smooth (0) or punctate throughout (1). *Character 30*: striae may be smooth (0) or punctate in lateral striae (1). *Character 31*: striae may be complete (0) or reduced to broken dashes (1). *Character 32*: elytral intervals may be moderately convex (0) or flat (1). *Character 33*: elytral intervals may be moderately convex (0) or costate (1). *Character 34*: elytral microsculpture may be an isodiametric scale (0) or a transverse mesh (1). *Character 35*: elytral microsculpture may be a transverse mesh (0) or dense transverse lines (1). *Character 36*: body color may be piceous (0) or metallic (1). *Character 37*: the elytral apex may be convex (0) or narrow, acuminate (1). *Character 38*: the elytral apex may be convex, rounded (0) or bearing a spine (1).

The Metathorax

Character 39: the metepisternum may have lateral edge longer than posterior edge (0), or lateral and posterior dimensions may be subequal (1). *Character 40*: flight wings may be fully developed (0), or the apex of the alae may be shortened (1). *Character 41*: the wings may possess a shortened apex (0), or the wings may be vestigial (1).

The Legs

Character 42: there may be a single lateral mesocoxal ridge seta (state 0, seta 23, Fig. 6-2B), or there may be more than one seta (1). *Character 43*: prothoracic fourth tarsomere with apical margin emarginate (state 0, as in Fig. 6-4A) or with apical margin bilobed (state 1, as in Figs. 6-4B–E). *Character 44*: metatrochanteral apex rounded (0) or acuminate (1). *Character 45*: metatrochanteral apex rounded (0) or emarginate (1). *Character 46*: tarsi on meso- and metalegs of male narrow (state 0, Figs. 6-4A–C) or expanded (state 1, Figs. 6-4D–E). *Character 47*: fourth tarsomere of metaleg with lobes of even length, to outer lobe twice as long as inner (state 0, Figs. 6-4A, E), or with outer lobe twice as long or more than length of the inner lobe (state 1, Figs. 6-4B–D). *Character 48*: mesofemur with 2–4 setae on ventral posterior surface (0), or with 5 or more setae (state 1, setae 24, Fig. 6-2B).

The Female Reproductive System

Character 49: apical gonocoxite with 2–3 lateral setae (0) or 4 or more lateral setae (1). *Character 50*: apical gonocoxite with normally proportioned lateral setae (0) or with lateral setae reduced to small pegs (1). *Character 51*: apical gonocoxite with dorsal seta present (0) or absent (1). *Character 52*: basal gonocoxite with apical setal fringe in a single row (0) or in at least a partially doubled row (1). *Character 53*: bursa copulatrix with field of microtrichia (0) or with microtrichia absent (1). *Character 54*: dorsal sclerite of bursa absent (0) or present (1).

The Male Genitalia

Derived character states are the presence of sclerotized spicules on the aedeagal internal sac, the regions defined in Fig. 6-6A. *Character 55*: sclerotized spicules at position I absent (0) or present (1). *Character 56*: sclerotized spicules at position II absent (0) or present (1). *Character 57*: spicules at position II short (0) or long (1). *Character 58*: sclerotized spicules at position III absent (0) or present (1). *Character 59*: spicules at position III short (0) or long (1). *Character 60*: sclerotized spicules at position IV absent (0) or present (1). *Character 61*: spicules at position IV short (0) or long (1). *Character 62*: sclerotized spicules at position V absent (0) or present (1). *Character 63*: spicules at position V short (0) or long (1). *Character 64*: sclerotized spicules at position VI absent (0) or present (1). *Character 65*: spicules at position VI short (0) or long (1). *Character 66*: internal sac as long as median shaft of aedeagus (0) or much longer than shaft (1).

Appendix 6.2: Derivation of Original Data Matrix

The taxon X character matrix could be derived from the character state advances and reversals shown on the cladograms (Figs. 6-7, 6-10, 6-11, 6-12, 6-13, 6-15), were it not for missing data for various characters in some taxa. This missing data is due to (1) the character state being precluded by derivation in another character or (2) lack of one or the other sex for the analysis.

To derive the taxon X character data matrix, use Figure 6-7 to obtain derived character states for each species group and for all species of the *baragua* group. Add derived states for the other species by consulting Figures 6-10, 6-11, 6-12, 6-13, and 6-15. Insert as missing data the following characters for the following taxa: Character 14, *P. acuniai, P. cubensis, P. cavicola, P. cinchonae, P. faber, P. jamaicae, P. rastafarius*; characters 49–54, *P. acuniai, P. constricticeps, P. laetificus, P. eupunctatus, P. parallelosomus, P. memnonius, P. paramemnonius, P. luciae*; characters 55–66, *P. cychrinus,*

P. elongatus, P. ellipticus, P. luciae. Fill in the remaining cells as the primitive character states.

Acknowledgments

I thank Alfred F. Newton, Jr., and James M. Carpenter of the Department of Entomology, Museum of Comparative Zoology, Harvard University, for lending me the taxonomic material of Philip J. Darlington, Jr. Discussion with other participants of this symposium, especially David A. Grimaldi and Stephen W. Nichols, substantially improved the manuscript. I thank George E. Ball and Gary R. Noonan for critical review of the manuscript, Kathleen Schmidt for the habitus drawing of *Platynus jaegeri*, and Susan Pohl and Velvet Saunders for typing and retyping the manuscript. Ann Hajek has been a source of encouragement and assistance from data gathering to reading proof. This project was supported by Hatch Project NY(C) 139406.

References

Anderson, T.H., and V.A. Schmidt. 1983. The evolution of Middle America and the Gulf of Mexico—Caribbean Sea during Mesozoic time. Geol. Soc. Amer. Bull. 94:941–966.

Arden, D.D., Jr. 1975. Geology of Jamaica and the Nicaraguan Rise. *In* A.E.M. Nairn and F.G. Stehli, eds., The Ocean Basins and Margins, vol. 3: The Gulf of Mexico and the Caribbean, pp. 617–661. Plenum Press, New York.

Baker, R.J., and H.H. Genoways. 1978. Zoogeography of Antillean bats. *In* F.B. Gill, ed., Zoogeography in the Caribbean, pp. 53–97. Spec. Publ. No. 13, Acad. Nat. Sci. Philadelphia.

Bils, W. 1976. Das Abdomenende weiblicher, terrestrisch lebender Adephaga Coleoptera und seine Bedeutung für die Phylogenie. Zoomorphol. 84:113–193.

Brown, F.M. 1978. The origins of the West Indian butterfly fauna, pp. 5–30. See Baker and Genoways 1978.

Buskirk, R.E. 1985. Zoogeographic patterns and tectonic history of Jamaica and the northern Caribbean. J. Biogeogr. 12:445–461.

Cadle, J.E. 1985. The Neotropical colubrid snake fauna (Serpentes: Colubridae): Lineage components and biogeography. Syst. Zool. 34:1–20.

Chace, F.A., Jr., and H.H. Hobbs, Jr. 1969. The freshwater and terrestrial decapod crustaceans of the West Indies with special reference to Dominica. U.S. Natl. Mus. Bull. No. 292, 258 pp.

Clench, H.K. 1963. A synopsis of the West Indian Lycaenidae with remarks on their zoogeography. J. Res. Lepid. 2:247–270.

Darlington, P.J., Jr. 1934. New West Indian Carabidae, with a list of the Cuban species, I. Psyche 41:66–131.

———. 1935. West Indian Carabidae, II: Itinerary of 1934; forests of Haiti; new species; and a new key to *Colpodes*. Psyche 42:167–215.

————. 1937a. West Indian Carabidae, III: New species and records from Cuba, with a brief discussion of the mountain fauna. Soc. Cubana Hist. Nat. 11:115–136.

————. 1937b. West Indian Carabidae, IV: Three new *Colpodes*. Psyche 44:122–124.

————. 1939. West Indian Carabidae, V: New forms from the Dominican Republic and Puerto Rico. Soc. Cubana Hist. Nat. 13:79–101.

————. 1943. Carabidae of mountains and islands: Data on the evolution of isolated faunas, and on atrophy of wings. Ecol. Monogr. 13:37–61.

————. 1953. West Indian Carabidae, IX: More about the Jamaican species. Occ. Papers Mus. Inst. Jamaica No. 8, 14 pp.

————. 1964. West Indian Carabidae, X: Three more species from Jamaica, including a new cave *Colpodes*. Psyche 71:181–182.

————. 1970. Carabidae on tropical islands, especially the West Indies. Biotropica 2:7–15.

Donnelly, T.W. 1985. Mesozoic and cenozoic plate evolution of the Caribbean region. *In* F.G. Stehli and S.D. Webb, eds., The Great American Interchange, pp. 89–121. Plenum Press, New York.

Fink, L.K., Jr. 1970. Field guide to the island of La Desirade with notes on the regional history and development of the Lesser Antilles Island Arc. *In* T.W. Donnelly, ed., International Field Institute Guidebook to the Caribbean Island-Arc System, Amer. Geol. Inst., Washington, D.C. 18 pp.

Flint, O.S., Jr. 1978. Probable origins of the West Indian Trichoptera and Odonata faunas. Proceedings of the 2d International Symposium on Trichoptera, pp. 215–223. Reading, England.

Freitag, R., J.E. Olynyk, and B.L. Barnes. 1980. Mating behavior and genitalic counterparts in tiger beetles (Carabidae: Cicindelinae). Intl. J. Invert. Reprod. 2:131–135.

Gorman, G.C., D.G. Buth, and J.S. Wyles. 1980. *Anolis* lizards of the eastern Caribbean: A case study in evolution, III: A cladistic analysis of albumin immunological data, and the definition of species groups. Syst. Zool. 29:143–158.

Hennig, W. 1966. Phylogenetic Systematics. University of Illinois Press, Urbana. 263 pp.

Lewis, J.F. 1980. Cenozoic tectonic evolution and sedimentation in Hispaniola, pp. 65–73. 9th Carib. Geol. Conf., Santo Domingo, Dominican Republic.

Liebherr, J.K. 1986a. *Barylaus*, new genus (Coleptera: Carabidae) endemic to the West Indies with Old World affinities. J. New York Entomol. Soc. 94:83–97.

————. 1986b. Cladistic analysis of North American Platynini and revision of the *Agonum extensicolle* species group. Univ. Cal. Publ. Entomol. 106:1–198.

————. 1987. A taxonomic revision of West Indian *Platynus* beetles (Coleoptera: Carabidae). Trans. Amer. Entomol. Soc. 112(1986):289–368.

Lindroth, C.H. 1969. The ground beetles of Canada and Alaska, part I. Opusc. Entomol. Suppl. 35. 47 pp.

————. 1974. On the elytral microsculpture of carabid beetles (Col. Carabidae). Entomol. Scand. 5:251–264.

Maurrasse, F.J.-M.R., F. Pierre-Louis, and J.G. Rigaud. 1980. Cenozoic facies distribution in the southern peninsula of Haiti and the Barahona Peninsula, Dominican Republic, pp. 161–174. 9th Carib. Geol. Conf., Santo Domingo, Dominican Republic.

Munroe, E.G. 1950. The systematics of *Calisto* (Lepidoptera, Satyrinae), with remarks on the evolutionary and zoogeographic significance of the genus. J. New York Entomol. Soc. 58:211–240.

Nelson, G., and N. Platnick. 1981. Systematics and Biogeography: Cladistics and Vicariance. Columbia University Press, New York. 565 pp.

Perkins, P.D. 1980. Aquatic beetles of the family Hydraenidae in the western hemisphere: Classification, biogeography and inferred phylogeny (Insecta: Coleoptera). Quaest. Entomol. 16:3–554.

Robinson, E., J.F. Lewis, and R.V. Cant. 1970. Field guide to aspects of the geology of Jamaica. *In* T.W. Donnelly, ed., International Field Institute Guidebook to the Caribbean Island-Arc System, 39 + 4 pp. Amer. Geol. Inst., Washington, D.C.

Rosen, D.E. 1975. A vicariance model of Caribbean biogeography. Syst. Zool. 24:431–464.

———. 1978. Vicariant patterns and historical explanation in biogeography. Syst. Zool. 27:159–188.

———. 1985. Geological hierarchies and biogeographic congruence in the Caribbean. Ann. Missouri Bot. Gard. 72:636–659.

Schwartz, A. 1978. Some aspects of the herpetogeography of the West Indies, pp. 31–51. See Baker and Genoways 1978.

Scott, J.A. 1972. Biogeography of Antillean butterflies. Biotropica 4:32–45.

Sober, E. 1983. Parsimony in systematics: Philosophical issues. Ann. Rev. Ecol. Syst. 14:335–357.

Southwood, T.R.E. 1962. Migration of terrestrial arthropods in relation to habitat. Biol. Rev. 37:171–214.

Steineck, P.L. 1974. Foraminiferal paleoecology of the Montpelier and lower coastal groups (Eocene-Miocene), Jamaica, West Indies. Paleogeogr., Paleoclimatol., Paleoecol. 16:217–242.

Swofford, D.L. 1984. P.A.U.P.: Phylogenetic Analysis Using Parsimony, version 2.3. Illinois Nat. Hist. Sur., Champaign.

Sykes, L.R., W.R. McCann, and A.L. Kafka. 1982. Motion of Caribbean Plate during last 7 million years and implications for earlier Cenozoic movements. J. Geophys. Res. 87 (B13):10656–10676.

Trueb, L., and M.J. Tyler. 1974. Systematics and evolution of the Greater Antillean hylid frogs. Occ. Pap. Mus. Nat. Hist. Univ. Kans., no. 24:1–60.

Wadge, G., and K. Burke. 1983. Neogene Caribbean Plate rotation and associated Central American tectonic evolution. Tectonics 2:633–643.

Watrous, L.E., and Q.D. Wheeler. 1981. The out-group comparison method of character analysis. Syst. Zool. 30:1–11.

Whitehead, D.R. 1973. Annotated key to *Platynus*, including *Mexisphodrus* and most "*Colpodes*," so far described from North America including Mexico (Coleoptera: Carabidae: Agonini). Quaest. Entomol. 11:591–619.

7 · Historical Biogeography of Two Groups of Caribbean *Polycentropus* (Trichoptera: Polycentropodidae)

Steven W. Hamilton

The order Trichoptera, or caddisflies, with more than 7000 described species, is relatively small when compared to the major insect orders such as Diptera, Lepidoptera, and Coleoptera. Conservative estimates predict as many as 10,000 species of caddisflies worldwide (Ross 1967), whereas more liberal estimates suggest numbers as high as 50,000 species (Schmid 1968, 1984). Such high estimates do not seem outlandish when considered in light of Erwin's (1983) estimate of 30 million insect species worldwide.

Caddis larvae are almost exclusively inhabitants of freshwater, occurring in a great variety of both lentic and lotic habitats. The larval stage is of particular interest to entomologists and ecologists because many species use their labial silk secretions for constructing cases, catchnets, or tubular re-treats (Wiggins 1977). Although generally less familiar, adult caddisflies are the basis for most current systematic schemes. The adults are drab, small-to-medium sized, hair-covered insects that somewhat resemble moths but lack scales and a coiled proboscis. Generally crepuscular or nocturnal, most adult Trichoptera can fly, and although some are quite strong flyers, most are relatively more fleet on foot than in flight (Davis 1934). Although a few species of caddis adults may fly considerable distances (up to 74 km) from freshwater habitats (Sparks et al. 1986), as a whole they tend to restrict their activities to the vicinity of their larval habitats, rarely ranging far afield (Davis 1934, Ross 1944, McCafferty 1981, Unzicker et al. 1982).

During the last two decades the intensive collecting and taxonomic activity of O. S. Flint, Jr., of the Smithsonian Institution, Washington, D.C., has contributed extensively to our knowledge of Neotropical caddisflies. In a presentation on the origins of the trichopteran and odonate fauna of the Antilles, Flint (1978) reported that over 80% of the Trichoptera in the Antilles are single-island endemics, which would seem to make this order

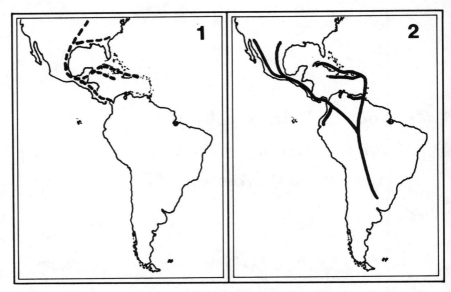

Figure 7-1, 7-2. Summary of transcontinental generalized tracks: 7-1, North American–Caribbean track; 7-2, South American–Caribbean track (after Rosen 1975, Figs. 6A and 6C).

ideal for study of Caribbean historical biogeography. Other findings of his study are summarized here.

1. The trichopteran fauna of the Greater Antilles is very distinct and has been isolated for a long time. There are three faunistic relationships: one with North America; the predominant one with the Neotropics, especially Central America; and the smallest one with Africa.
2. Although distinct, the Lesser Antillean Trichoptera fauna has a lower percentage of endemism and shows relationships primarily with northern South America. A small percentage of this insular fauna is shared with the mainland of South America.
3. ". . . microcaddisflies are more widely distributed than average sized caddisflies" and this may be because the small caddisflies are more readily blown about. There is also an ecological relationship whereby "those species breeding in warm, lower elevation ponds or slow-flowing streams are more widely dispersed than those requiring cooler, faster-flowing upland brooks (pp. 222–223)."
4. One should be cautious in "applying the results of analysis of one group to another group." Even though 80% of the Trichoptera are single-island endemics, compared to only 15% of the Odonata, Flint believes "that the basic origins of the fauna of these two orders is the same (p. 223)."
5. The trichopteran and odonate faunal distributions are "in basic agreement with the vicariance model of Caribbean biogeography (p. 223)" advanced in 1976 by Rosen (Figs. 7-1 and 7-2).

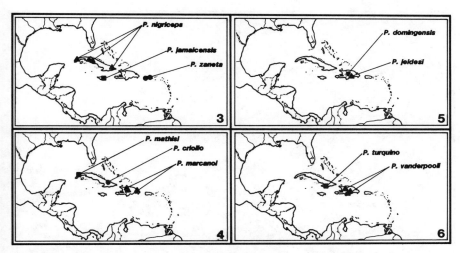

Figure 7-3 to 7-6. Species distributions in the *Polycentropus nigriceps* group.

Flint (1976) made a detailed study of the phylogeny and historical bio-geography of the *Polycentropus nigriceps* group, a monophyletic species group then comprising seven species, each endemic to single islands in the Greater Antilles: *P. nigriceps* Banks from Cuba; *P. jamaicensis* Flint from Jamaica; *P. zaneta* Denning from Puerto Rico; and *P. domingensis* Banks, *P. jeldesi* Flint, *P. marcanoi* Flint, and *P. vanderpooli* Flint from Hispaniola. Subsequently, three Cuban species allied to this group have been described: *P. criollo* and *P. turquino* by Botosaneanu (1980), and *P. mathisi* by Hamilton (1986a). A majority of this study pertains to these 10 species (Figs. 7-3–7-6). My conclusions agree with Flint's (1976) hypothesis that the *confusus* group, with 15 species in eastern North America, is the sister taxon to the *nigriceps* group and that these two groups represent components of the North American–Caribbean track (Fig. 7-1). Our hypotheses of the cladistic relationships within the *nigriceps* group are in partial disagreement.

The second species group considered below, the *Polycentropus insularis* group (Flint 1974), is a smaller, monophyletic taxon of 5 species. One species, *P. insularis* Banks, occurs in the Lesser Antilles, while the other four species, *P. altmani* Yamamoto, *P. biappendiculatus* Flint, *P. cuspidatus* Flint, and *P. surinamensis* Flint are variously distributed in northern South America and southern Central America (Fig. 7-7). Below, I present a hypothesis of the sister-group relationship between the *insularis* group and the large, Neotropical *gertschi* group; I suggest that these groups represent components of the South American–Caribbean track (Fig. 7-2).

Most current studies of historical biogeography use methods of vicariance biogeography that have been discussed frequently in recent years (Croizat et

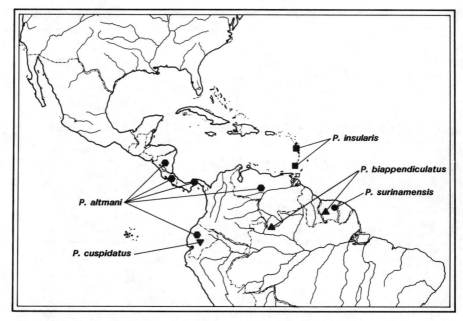

Figure 7-7. Species distributions in the *Polycentropus insularis* group.

al. 1974, Platnick and Nelson 1978, Nelson and Rosen 1981, Nelson and Platnick 1981). Ichthyologist D. E. Rosen of the American Museum of Natural History has specialized in Mesoamerican historical biogeography, and has made contributions both theoretically (Rosen 1975, 1978, 1985) and practically (Rosen 1974, 1979, and others). Croizat, Nelson, Platnick, and Rosen have been the major forces in placing historical biogeography into a scientific framework (Humphries 1982).

The biological complexity in the Caribbean region is partially a reflection of the complexity of the geologic history of the region. Models describing Caribbean geohistory are numerous (Malfait and Dinkleman 1972, Donnelly 1975, Pindell and Dewey 1982; see also Hedges 1982, Donnelly 1985, and Rosen 1985 for more complete summaries of this literature), and they are of two general types—stabilist and mobilist. The stabilist models hypothesize that the islands of the Caribbean formed in essentially their current positions. Alternatively, mobilist models hypothesize that the Caribbean islands originated in either the eastern Pacific Basin or the southern Caribbean Basin and then moved, via the forces of plate tectonics, to their present positions. Within each of these two camps, several secondary models exist, each differing to various degrees (ages, vectors, rates, data used, data ignored, etc.). Mobilist models of Caribbean geohistory have resulted in several vicariance models predicting the regional biohistory (Rosen 1975, 1978, 1985, Flint 1976, Savage 1982).

With so many diverse geologic models to explain the history of one small region of the earth, historical biogeographers are left to choose among geologic models supported by types of data with which they have little familiarity. In reality, systematists and historical biogeographers do not need to choose a particular geohistorical model, for they have their own methods of estimating biohistory. That method is the construction of a cladogram of hierarchically nested homologues that predicts the relationship of the taxa and the areas in which they occur. Of course, a single cladogram predicts only the relationship of a single group of organisms, and by itself it is a rather weak hypothesis for predicting relationships of geographic areas. Congruent patterns of area relationships based on cladistic analyses of numerous taxa suggest a common cause. That is, if we find that cladistic analysis of many varied monophyletic groups of plants and animals demonstrates the same cladistic relationships of geographic areas, then it is more parsimonious to ascribe that shared distributional pattern to a single historical event that affected the entire ancestral biota, rather than to numerous independent events. Because the pattern of area relationships predicted from biological data will not reveal whether the ancestral biota vicariated or dispersed en masse (biotic dispersal), the systematist must examine the geologic literature for evidence of the geohistorical area relationships (Platnick and Nelson 1978). At the same time, geologists can independently search for the geohistorical data and generate geologically derived area cladograms (Rosen 1978). With such reciprocal illumination between the fields of biology and geology, the future of historical biogeography looks very bright.

Character Analysis

The diagnosis and recognition of trichopteran species is based almost exclusively on the structures of male terminalia, specifically those structures associated with mating (Fig. 7-8). The terminology used here follows closely that used by Schmid (1980). As in all polycentropodid caddisflies, and most Trichoptera in general, abdominal segments I–VIII of *Polycentropus* are unmodified, although sternite and tergite VIII are sometimes narrower than on preceding segments. Sternite IX (s. IX) is generally large and heavily sclerotized, being broad ventrally and narrowed and obsolete at one-half to three-quarters of the height of the abdomen. The unsclerotized tergite IX and segment X (t. IX + X) are apparently united to form the dorsoapical body, which bears the anus. This body is sometimes partially sclerotized below and bears ventrally paired, sclerotized processes. The inferior appendages (inf. app.), or claspers, are paired, movable structures located ventrally on sternite

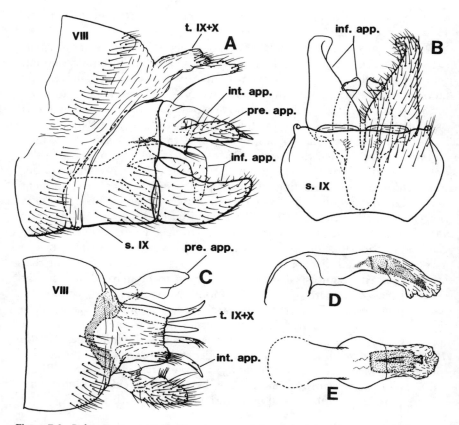

Figure 7-8. *Polycentropus criollo* Botosaneanu: inf. app. = left inferior appendage; int. app. = left intermediate appendage; pre. app. = left preanal appendage; VIII = abdominal segment VIII; s. IX = sternite of abdominal segment IX; t. IX + X = fused terga of abdominal segments IX and X.

IX. The paired preanal appendages (pre. app.), or cerci, articulate with the posterodorsal corners of sternite IX. The preanal appendages are generally setose, earlike processes but are frequently modified, bearing dorsal, mesal, or ventral sclerotized branches. The intermediate appendages (int. app.) are normally thin, elongate processes fused or articulating basally with the preanal appendages. The phallus (Fig. 7-8D and E) is a tubular structure protruding from the usually membranous phallocrypt that lies below the membranous segment IX + X, between the preanal appendages, and above the united bases of the inferior appendages. The phallus is a tubular structure that is sclerotized basally and membranous apically and may or may not have accessory spines ranging from one to greater than ten in number. Embedded

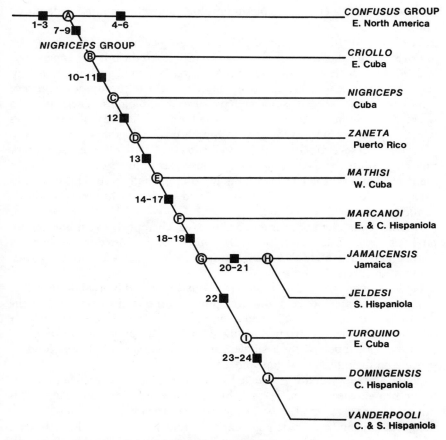

Figure 7-9. Cladogram of the *Polycentropus nigriceps* group.

in the eversible membrane is a large sclerite (phallotremal sclerite ?), which is apical when the membrane is everted.

The *Nigriceps* Group (Fig. 7-9)

Monophyly of the *confusus* group plus *nigriceps* group (node A), a relationship hypothesized by Flint (1976), is indicated by three homologues (= synapomorphies). The strongest of the three consists of (1) the paired, ventrally positioned, sclerotized processes on the otherwise membranous segment x. The other two homologues, (2) the dorsobasal process on the inferior appendage and (3) the phallic apparatus reduced to the phallobase with a large, potentially extrudible phallotremal sclerite, may not be exclusive to this unit (they may be plesiomorphic, i.e., define a more generalized unit).

The (4) elongation of the ventral process of segment x, (5) the rugose, darkly sclerotized point on the dorsobasal process of the inferior appendage, and (6) the reduction of the size of the preanal appendage all imply monophyly for the *confusus* group. Monophyly of its sister group, the *nigriceps* group (node B), is also supported by three homologues. There is (7) a sclerotized subphallic bridge between the bases of the preanal appendages which appears to support the phallic apparatus. Analogous subphallic supports occur in several groups of *Polycentropus s. l.* Other homologues of the *nigriceps* group are (8) the paired setiferous lobes or areas on the dorsum of segment x, and (9) the flat, lateral flanges on the phallobase. This character (9) appears to be lost in the two most derived species, *P. domingensis* and *P. vanderpooli.*

Another character not mentioned above, which appears to support monophyly of the *nigriceps* group (node B), consists of the paired lateral flaps on sternite IX of the female abdomen. Unfortunately, females are not yet associated for all species of this group.

Node C (*nigriceps* group minus *criollo*) is supported by two homologues on the inferior appendage, (10) development of the vertical brace below the dorsobasal process, and (11) reduction of the basal apodeme. Monophyly for the group arising from node D (*nigriceps* group minus *criollo* and *nigriceps*) is supported by a single, unique homologue, (12) a secondary process on the intermediate appendage.

Node E is supported by a single homologue, (13) the presence of a posterior prominence on the vertical brace of the inferior appendage. This prominence is very small in *P. jamaicensis* and *P. marcanoi.* Monophyly of the group arising from node F is clearly demonstrated by four homologues. Most apparent is (14) the shortened, thickened, specialized setae on the preanal appendage. Also, (15) the preanal appendage is greatly elongated, (16) the vertical distance between the preanal and intermediate appendages is greatly reduced, and there is (17) a patch of setae on a small bump on the mesal surface of the vertical brace of the inferior appendage. Two homologues supporting node G are (18) the strong dorsad curve at the base of the preanal appendage, and (19) the strong ventrad curvature of the dorsobasal process of the inferior appendage.

The sister relationship of *P. jamaicensis* and *P. jeldesi* (node H) is supported by two characters of the preanal appendage, (20) the upturned apex, and (21) the lateral ridge that is strongly expressed laterad. Monophyly of their sister unit (*P. turquino*, *P. domingensis*, and *P. vanderpooli*), indicated by node I, is supported by a single homologue, (22) separation of the preanal appendage into two parts: one lateral and bearing normal, thin setae, and the other more apicomesal and bearing only specialized setae.

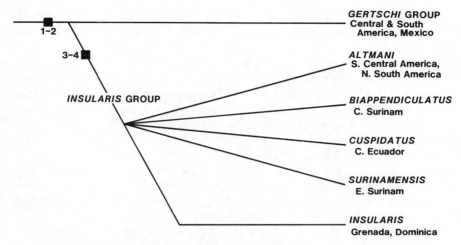

GERTSCHI GROUP
Central & South
America, Mexico

1-2

3-4

ALTMANI
S. Central America,
N. South America

INSULARIS GROUP

BIAPPENDICULATUS
C. Surinam

CUSPIDATUS
C. Ecuador

SURINAMENSIS
E. Surinam

INSULARIS
Grenada, Dominica

Figure 7-10. Cladogram of the *Polycentropus insularis* group.

The sister relationship of *P. domingensis* and *P. vanderpooli* (node J) is indicated by two homologues. One is the further expression of homologue 22, the (23) separation of the preanal appendage into two parts. The extreme of character 23 is in *P. vanderpooli*, where the preanal appendage is divided into two distinct lobes; the more mesal lobe is covered with specialized setae and the lateral lobe resembles the more plesimorphic form and position of the preanal appendage (as in *P. nigriceps*). The other character supporting node J is (24) the reduction of the lateral flange on the phallobase.

The *Insularis* Group (Fig. 7-10)

I have used the *gertschi* group as the out-group in my analysis of the *insularis* group. Flint (1980) provided several characters to define his *gertschi* group but admitted that some species would violate some of the definition. His first character for the *gertschi* group is inferior appendage (clasper) "a thin, erect dorsolateral lobe that joins a thicker ventromesal region which is delimited by a sharp mesal shelf or carina that bears one or two sharp toothlike projections" (1980:149). This is a good gestalt character that is violated variously in numerous species and is weak at best. Flint's second character is that the preanal appendage (cercus) generally has three lobes, "a slender rodlike dorsomesal lobe [= intermediate appendage], a broad usually elongate or quadrate dorsolateral lobe (densely setate), and a ventral and slightly more mesal lobe usually broadly joined to the dorsolateral lobe" (1980:149–151). Although again having a useful gestalt value, this character and its individual components are not unique. The general form of this dorsomesal lobe (intermediate appendage) is shared by the *confusus* group, *nigriceps* group, *arizonensis* group (Flint 1980), and *insularis* group. The dorsolateral lobe (= preanal appendage) is equally widespread in these groups. The other mesal lobe, which is usually broadly joined to the dorsolateral lobe, is very widespread in the Polycentropodidae and as such is useless at this level of the analysis. I find the monophyly of the *gertschi* group as currently defined unconvincing and believe it may involve a larger group, including the *insularis* group and possibly the *arizonensis* group. It is not necessary to resolve these questions here. I have presented a phylogeny of the *insularis* group elsewhere (Hamilton 1987).

Monophyly of the *gertschi* group plus *insularis* group is demonstrated by two homologues. The (1) apicomesal lip of the phallobase is attenuated into a long, acute, ventral process, and (2) the narrow intermediate appendage has a distinct flexion point, or articulation, usually near its base.

Monophyly of the *insularis* group is supported by two homologues. The (3) intermediate appendage extends anteriorly into the abdomen below the membranous segment x and above the phallus before being recurved mesally, then posteriorly, and exceeding the rest of the genitalia. The flexion point on the intermediate appendage is at the recurved section. The second homologue is (4) the pair of heavy endophallic processes, which are dorsally positioned on the retracted phallic apparatus.

Biogeographic Conclusions

In the following discussion my results will be compared with the models and predictions of Rosen (1975, 1985) because they are the only source of area cladograms for the Caribbean region. Rosen's 1975 models, which deal with the relationship of the Antilles to adjacent continental land masses, are based both on biogeographic data (from numerous sources) and on geologic data (primarily Malfait and Dinkleman 1972). His most recent models (Rosen 1985) predict inter— and extra—Greater Antillean relationships from several mobilist models of Caribbean geohistory and use no biogeographic data. Of course, Rosen's predictions are based on vicariance biohistorical models that are necessarily linked to mobilist models of Caribbean geohistory. This approach may be to the disfavor of proponents of stabilist models, but no similar biohistorical models based on a stable geohistory are available. Donnelly (1985:115) recently compiled a large bank of data supporting a stabilist model of Caribbean geohistory, which, he says "provides a framework upon which biogeographic hypothesis can be erected and invites biogeographic tests of its validity." Because his model provides no evidence of past dry-land connections between the Antilles and any of the continental landmasses, the ancestors of the current Antillean biota must have reached the islands strictly via dispersal. Unfortunately, models that use past dispersal to explain current biotic distributions are untestable (Nelson and Platnick 1981). Dispersal can be used to explain any distribution, and, as Wiley (1981:287) so clearly stated, "dispersal, in explaining everything, really explains nothing." The practical and philosophical difficulties entailed in dispersalist models have been discussed in detail by several authors in recent years (Croizat et al. 1974, Platnick 1976, Nelson and Platnick 1981, Nelson and Rosen 1981, Wiley 1981, and others). Rosen (1975) has addressed these difficulties in general and as they specifically relate to Caribbean biogeography. Readers who still cling to the notion of using dispersal as the first and primary mechanism for explaining biogeographic distributions should consider the arguments presented in these papers and books. This is not a denial that dispersal is a common occurrence, but for methodological reasons dispersal should be a last resort for explaining modern distributions and used only after all vicariance possibilities have been considered. In fact, the strongest biogeographic support for stabilist models of Caribbean geohistory would come from a lack of sound evidence of vicariance, which is, unfortunately, negative data of little scientific value. In this case the geologic evidence would be of overwhelming importance. Ultimately, if geologic evidence convincingly supports the stabilists' models over the mobilists' models, then the only explanation for the various generalized biogeographic tracks involving the Antilles is dispersal. In fact, based on geologic evidence indicating that the Lesser Antilles formed in position as the result of very recent (40 Ma or less) volcanic activity (Coney 1983), the Antillean portion of the South American–Caribbean track can be explained only by dispersal of northern South American ancestors. Based on the geologic data, Rosen's (1975) vicariance explanation for this portion of the SA–C track must be rejected.

Although I will compare my results to Rosen's models, I believe more cladistic analyses of Caribbean taxa are currently needed to discover the most highly corroborated generalized tracks. This type of analysis is needed desparately for groups endemic to the Greater Antilles in order to provide intra-Antillean tracks. Once they are known, dispersal should be apparent as individual or minimally corroborated tracks and the established generalized tracks will provide substantial data for consideration of geologic models.

Rosen (1975) presented data supporting two New World generalized tracks that involve the Antilles, the North American–Caribbean (NA–C) track and the South American–Caribbean (SA–C) track (Figs. 7-1 and 7-2). Among the species of *Polycentropus*, two monophyletic groups with Antillean members appear to fit Rosen's models. The monophyletic unit of the *confusus* + *nigriceps* groups comprises major components of the highly corroborated NA–C track, but it lacks the Mexican–Nuclear Central American component apparent in so many other groups. Rosen (1975) has argued that it is more parsimonious to include groups with missing components in this major track rather than to hypothesize a second NA–C track via Florida. Of course, discovery of a member of the *confusus* + *nigriceps* group in Mexico or Nuclear Central America (or Florida) would provide strong evidence of track membership.

Components of the SA–C track, also a highly corroborated track, are found in the *insularis* group. As stated above, based on current geologic evidence this portion of the track must be attributed to dispersal. This species group is a member of a larger, unnamed, monophyletic unit that includes most Neotropical *Polycentropus*. With *P. insularis* on Grenada and Dominica, and with the remaining four species in northern South America and southern Central America, the *insularis* group comprises the middle component of the larger SA–C track. A complex of undescribed species from southeastern Brazil may represent a southern component of this track, and the *gertschi* group may be a more northern component. The cladistic relationships of this large Neotropical complex of *Polycentropus* are currently under study (Hamilton 1986b).

Because each of the 10 species in the *nigriceps* group is a single-island endemic in the Greater Antilles, little in my phylogeny can be fit to any of Rosen's (1985) series of four-area cladograms. The monophyletic unit arising from node G loosely fits one four-area cladogram derived from a "consensus map of the mid-Cenozoic relations among some main components of the Caribbean heartland" (Rosen 1985:652, Figs. 35 and 36). One of these four-area cladograms indicates sister-area relationships of Jamaica with southwest Hispaniola (hypothesized by some geologists as a once distinct area) and of eastern Cuba (hypothesized as a once distinct area) with central Hispaniola. This is generally the relationship of *P. jamaicensis*, *P. jeldesi*, *P. turquino*, and *P. domingensis* + *P. vanderpooli*. *Polycentropus jamaicensis* from Jamaica is the sister species to *P. jeldesi*, which is only known from central Hispaniola, and *P. turquino* from eastern Cuba is the sister species to *P. domingensis* plus *P. vanderpooli*, both from central and south-central Hispaniola.

Within the *nigriceps* group, no other clades appear to fit any of Rosen's (1985) area cladograms. A character analysis of the *nigriceps* group based on the semaphoront (not just the adult male) and more complete distributional data (especially from Cuba and Haiti) would undoubtedly test and enhance the results I have presented here. This more in-depth analysis might result in other patterns which could be compared to the models of Rosen and others. Does the lack of a good fit of the *nigriceps* group cladogram to Rosen's area cladograms indicate interisland dispersal? Not necessarily. A single group does not provide the quantity of evidence needed to test biogeographic and geologic models. This evidence will come only from the study of many more groups of organisms.

Summary

A recent study of Antillean Trichoptera (Flint 1978) indicated a high percentage (80%) of single-island endemism in the Greater Antilles and a much lower percentage in the Lesser Antilles. The Greater Antillean Trichoptera showed faunistic relations to North America, Central America, and Africa. The Lesser Antillean caddisflies had affinities to northern South America. These patterns are in agreement with two terrestrial, generalized tracks documented by Rosen (1975), a North American–Caribbean track and a South American–Caribbean track.

Two species groups of New World *Polycentropus*, the *nigriceps* group and the *insularis* group, are faunistically related to the Antilles. The *nigriceps* group comprises 10 species that are endemic to the Greater Antilles. The sister relationship of this group to the *P. confusus* group of eastern North America was hypothesized by Flint (1976) and is supported by the cladistic analysis presented here. This sister-group relationship is hypothesized to be part of Rosen's North American–Caribbean track. Rosen's vicariance model of Caribbean biogeography predicts that this pattern is the result of crustal plate movements hypothesized by several mobilist models of Caribbean geologic history. In these models the Greater Antilles and their associated biota drifted eastward from a position adjacent to nuclear Central America. Recent stabilist models of Caribbean geohistory hypothesize that the Greater and Lesser Antilles have formed in essentially their current position, never having been in contact with adjacent continental landmasses. If this is the case, then the generalized Caribbean track must be the result of dispersal from the adjacent continents.

The cladistic analysis of the 10 species of the *nigriceps* group gives no clear evidence of distinct intraisland patterns of relationship. Before sound hypotheses of interisland relationships can be proposed, other Greater Antillean groups must be cladistically analyzed and compared to find if a common pattern exists.

The *insularis* group has four species variously distributed in northern South America and southern Central America, and one species, *P. insularis*, is found on Grenada and Dominica. This group fits into Rosen's South American–Caribbean track. Current geologic evidence shows that the Lesser Antilles are of recent volcanic origin and have not drifted into their present position from elsewhere. Therefore, the biota of the Lesser Antilles must be the result of dispersal from northern South America, and not of vicariance as initially proposed by Rosen (1975).

I support the continued effort of systematists in conducting careful and detailed cladistic analyses of organisms from in and around the Caribbean region, which may result in the generation of more detailed patterns of area relationships. Vicariance explanations should be considered before dispersal scenarios are constructed. Of course, these hypotheses are limited by what is known of the geologic history of the region. The multiplicity and diversity of geohistorical models for the Caribbean region may leave historical bio-geographers confused. In this case, the biogeographer has cladograms to estimate the biological and geologic history of the region.

Acknowledgments

For their constructive comments on various stages of this manuscript I am indebted to R. T. Schuh of the American Museum of Natural History, P. D. Adler, R. W. Holzenthal (now of the University of Minnesota), R. R. Montanucci, J. C. Morse, and M. Rothschild of Clemson University, J. A. Slater of the University of Connecticut, O. S. Flint, Jr., and D. B. Wahl (now of the American Entomological Institute) of the Smithsonian Institution, and J. K. Liebherr of Cornell University. These individuals have contributed significantly to the clarity of this chapter. Any remaining lack of clarity is, of course, my fault.

This is Technical Contribution No. 2599 of the South Carolina Agricultural Experiment Station, Clemson University, Clemson, South Carolina.

Appendix: Taxonomic Data

In illustrations of the male terminalia in Figures 7-8 and 7-11 through 7-21, A, B, and C are lateral, ventral, and dorsal aspects, respectively, and D and E are lateral and dorsal aspects, respectively, of the phallic apparatus. To avoid confusion, all paired structures of the male genitalia, including the inferior appendages, the preanal appendages, the intermediate appendages, and the ventral processes of segment x, are discussed in the singular. Species-distribution data are based on published records and material I have examined. Specimens examined in this study were borrowed from the Canadian National Collection, Ottawa; the Illinois Natural History Survey, Urbana; the Museum of Comparative Zoology, Cambridge, Massachusetts; and the National Museum of Natural History, Washington, D.C.

Polycentropus Curtis

Polycentropus Curtis, 1835: plate 544. (Type-species: *Polycentropus irroratus* Curtis, 1835, by original designation.)

The genus *Polycentropus s.l.* comprises about 200 species worldwide and is most diverse in the Northern Hemisphere, although the Neotropical Region has a rich fauna. Many workers outside North America divide the species of *Polycentropus s.l.* among three genera—*Holocentropus, Plectroc-*

nemia, and *Polycentropus s.s.* All of the Neotropical species would be included in *Polycentropus s.s.*

Schmid (1980) has characterized the adults of *Polycentropus s.l.* and differentiates the three more restricted genera. A brief description of the larval morphology and biology is provided by Wiggins (1977). Lepneva (1964) gives a more detailed description of the morphology and biology of the larvae and pupae of each of the three genera.

The *Nigriceps* Group

Polycentropus nigriceps group, Flint, 1976:234.

This monophyletic group of 10 species is endemic to the Greater Antilles. Flint (1976) named this group and provided a diagnosis. This group is further characterized by the cladogram provided in Figure 7-9. Females are associated with males in 5 species (*P. criollo, P. jamaicensis, P. marcanoi, P. zaneta,* and *P. mathisi*), and two other female morphotypes remain unassociated (Flint 1976, Botosaneanu 1980, Hamilton 1986a). Larvae have been associated for only *P. jamaicensis* (Flint 1968a) and *P. zaneta* (Flint 1964). The scant biological data for species of the *nigriceps* group indicate an association with small mountain and hill streams (Flint 1976).

Synonymies and short diagnoses for males of each species are provided below. Because of the lack of associated immature and female material, only males were used for the character analysis. Also included for each species are summaries of the known distributions and biological data when available. Measurements of male forewing length were taken only from specimens that I had on hand.

Polycentropus criollo Botosaneanu

Polycentropus criollo Botosaneanu, 1978:229. *Nomen nudum.*
Polycentropus criollo Botosaneanu, 1980:101–102, Fig. 5; Hamilton, 1986b:56, Fig. 3.2.

Although *P. criollo* (Fig. 7-8) was placed in the *confusus* group by Botosaneanu, Flint, who examined the figures of this species, correctly surmised that it is the least derived member of the *nigriceps* group. Much of what is unique about this species results from this relatively underived state. Each preanal appendage is thumblike, and the intermediate appendage is short, curved laterad, and only slightly exceeds the preanal appendage. The lateral and ventral surfaces of segment x are sclerotized and bear a pair of ventral digitate processes. The vertical brace of the inferior appendage is absent and the dorsomesal process is curved downward at a right angle. The anterior apodeme of the inferior appendage is well developed. The phallobase has the lateral flanges typical of the *nigriceps* group. Male forewing length: 6.3 mm (*n* = 1).

Distribution. This species is known only from the holotype collected at

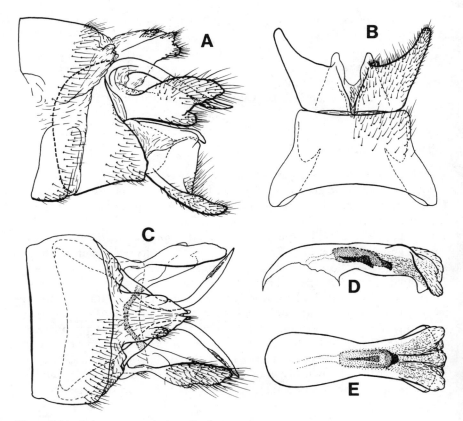

Figure 7-11. *Polycentropus nigriceps* Banks.

Pinares de Mayari, Prov. Oriente (Botosaneanu 1980). No more details of the collection locality were available, but according to Botosaneanu (1980) the Pinares de Mayari is a vast region of forested mountains, apparently in the Sierra de Nipe of Cuba (Fig. 7-4).

Polycentropus nigriceps Banks
Polycentropus nigriceps Banks, 1938:301, Fig. 5; Fischer, 1962:90; Flint, 1967b:6, Fig. 22, 1968b:80, 1976:235, Figs. 1–2; Botosaneanu, 1978:229, 1980:95; Hamilton, 1986b:63–64, Fig. 3.8.
Polycentropus rosarius Kingsolver, 1964:257, Figs. 1A–C; Flint, 1976:235 (as jr. syn. of *P. nigriceps* Banks).

P. nigriceps (Fig. 7-11) is readily recognized by two plesiomorphic conditions, the lack of both the secondary process of the intermediate appendage and the specialized setae on the short, preanal appendage. The inferior

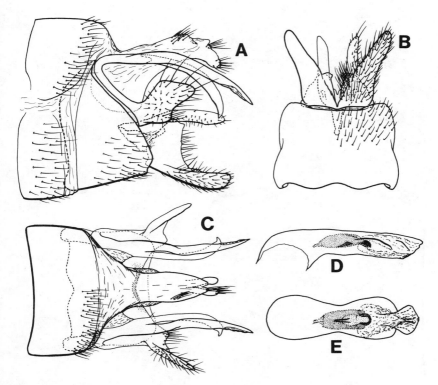

Figure 7-12. *Polycentropus zaneta* Denning.

appendage's vertical brace supports a small dorsobasal process. Although not associated with certainty, Flint (1976, Fig. 15) believes his "species 1" may be the female of *P. nigriceps*. Male forewing length: 7.3–7.7 mm (*n* = 4).

Distribution. Widespread in Cuba. Specimens have been collected from mountainous areas of the eastern Oriente Province, the central Las Villas Province, and the western Pinar del Rio Province (Fig. 7-3).

Polycentropus zaneta Denning

Polycentropus zaneta Denning, 1947:660, Figs. 4, 4A; Wolcott, 1950:93; Flint, 1964:32–33, Figs. 5F–G, 8I–L, 1968b:80; Fischer, 1972:35; Flint, 1976:235–236, Figs. 3–4, 16; Hamilton, 1986b:66–67, Fig. 3.11.

The secondary process on the intermediate appendage of this species (Fig. 7-12) is longer and thicker than in the other species. The preanal appendage is similar in size and shape to that of *P. nigriceps* and *P. mathisi*. The dorsobasal process of the inferior appendage is relatively straight and long

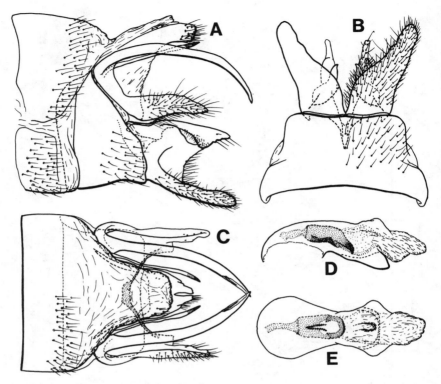

Figure 7-13. *Polycentropus mathisi* Hamilton.

compared to its vertical brace and to this process in other species of the group. Male forewing length: 5.5–6.4 mm (*n* = 4).

Distribution. Flint (1976) reported this species "from the mountains in the center and west of the island" of Puerto Rico (Fig. 7-3). It has also been recorded from the eastern end of the island (Denning 1947).

Biological data. The type locality of *P. zaneta* was described as a "roaring stream coming out of the high tropical rain forest" at an elevation of about 150 m above sea level (Denning 1947:660). Flint (1964) reported the habitat of the immatures as "pools and other slowly flowing areas in clear mountain streams."

Polycentropus mathisi Hamilton
Polycentropus mathisi Hamilton, 1986a:731–733, Figs. 1–6; 1986b:62, Fig. 3.7.

The preanal appendage of this species (Fig. 7-13) is of medium length and acute apically. The secondary process of the intermediate appendage is

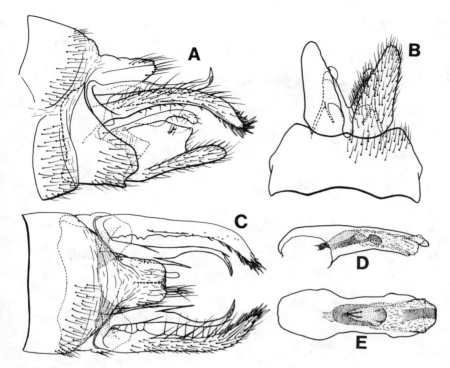

Figure 7-14. *Polycentropus marcanoi* Flint.

elongate, thin, lightly sclerotized except at its apex, and originates meso-basad on the intermediate appendage. The dorsobasal branch of the inferior appendage is broadened and almost hatchet-shaped apically. The vertical brace bears a large, upcurved prominence posteriorly. The phallobase has broad lateral flanges and an apicoventral lip. Male forewing length: 6.2 mm ($n = 1$).

Distribution. This species is only known from the type locality, Soroa, Pinar del Rio Province in western Cuba (Fig. 7-4).

Polycentropus marcanoi Flint

Polycentropus marcanoi Flint, 1976:238–239, Figs. 11–12, 18; Hamilton, 1986b:61–62, Fig. 3.6.

The preanal appendage of *P. marcanoi* (Fig. 7-14) bears on its apicomesal surface short, specialized setae. It is evenly curved in lateral and dorsal aspects, and is not subdivided. The intermediate appendage bears an acute secondary process of less than one-half its total length. The inferior appendage differs from that of other species in its general form, especially in the shape of the vertical brace, which bears an indistinct posterior prominence.

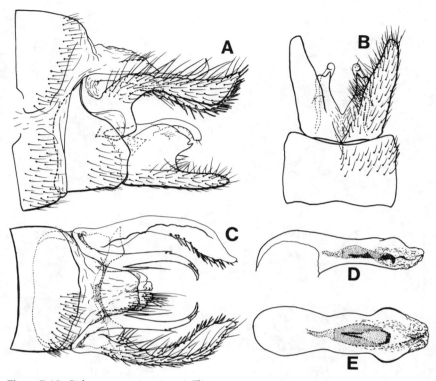

Figure 7-15. *Polycentropus jamaicensis* Flint.

The ventral lobe on the sclerotized venter of segment x is longer and more prominent than that in other species, but similar to that on *P. turquino*. The phallobase is moderately compressed with lateral flanges, and there is an infolded, apical, sclerotized band. Male forewing length: 5.5–6.4 mm (*n* = 5).

Distribution. This species is known only from the types collected in elevated areas of central and eastern Hispaniola (Fig. 7-4).

Polycentropus jamaicensis Flint
Polycentropus jamaicensis Flint, 1968a:25, Figs. 32–34; 1968b:80; 1976: 239, Figs. 13–14, 17; Hamilton, 1986b:58–59, Fig. 3.4.

Males of *P. jamaicensis* (Fig. 7-15) can be differentiated from those of other species by the strongly sinuate, apically incurved, and upturned preanal appendage. There is further differentiation of this appendage: it bears a strong lateral ridge, the lateral surface bearing long, thin setae, and the ventral and apicomesal surface bearing shortened, specialized setae. The

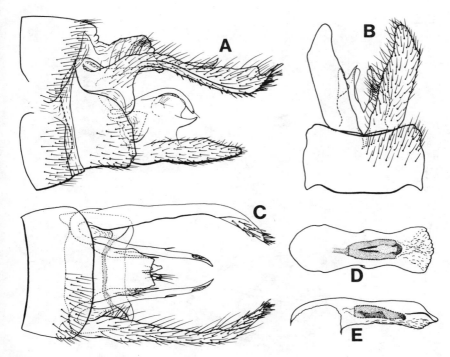

Figure 7-16. *Polycentropus jeldesi* Flint.

secondary process of the intermediate appendage is similar to that in *P. marcanoi*. The vertical brace of the inferior appendage has a small hooklike posterior prominence, and the dorsobasal process is thin and evenly but strongly curved ventrad. Male forewing length: 8.0–8.3 mm (*n* = 2).

Distribution. This species is known from a few examples collected in the mountains of eastern Jamaica (Fig. 7-3).

Biological data. "The larvae were taken in slowly flowing sections of a small mountain stream only a few feet in width. They were found in silken nets under flat stones" (Flint 1968a:25).

Polycentropus jeldesi Flint
Polycentropus jeldesi Flint, 1976:237–238, Figs. 7-8; Hamilton, 1986b:60, Fig. 3.5.

The preanal appendage of *P. jeldesi* (Fig. 7-16) is long and strongly sinuate, more sharply curved at the base, and narrower along its length than in *P. jamaicensis*, its sister species. The secondary process of the intermediate appendage is like that of *P. jamaicensis*. The ventrolateral branch has a lateral ridge, is notched and incurved near the apex, and bears the typical,

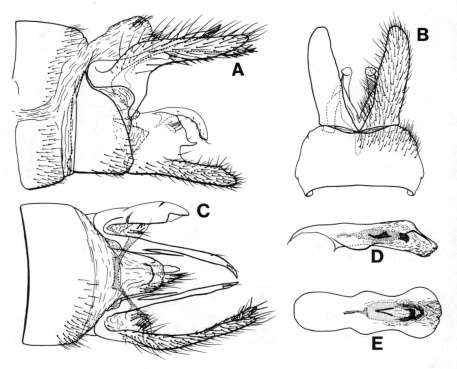

Figure 7-17. *Polycentropus turquino* Botosaneanu.

specialized setae mesoapically. The acute prominence on the posterior margin of the inferior appendage's vertical brace is much larger than in *P. jamaicensis*, but the dorsobasal process is very similar in these two species. Flint (1976, Fig. 19) believes that his "species 2" may be the female of *P. jeldesi*. Male forewing length: 8 mm (Flint 1976).

Distribution. This species is known only from the holotype taken in the mountains of the south-central Dominican Republic (Fig. 7-5).

Polycentropus turquino Botosaneanu
Polycentropus turquino Botosaneanu, 1978:229. *Nomen nudum.*
Polycentropus turquino Botosoneanu, 1980:102–104, Fig. 6; Hamilton, 1986b:64–65, Fig. 3.9.

The shape of the preanal appendage in *P. turquino* (Fig. 7-17) is similar to that in *P. jeldesi*, but in the former it is shorter, not sinuate apically, and the apex is differentiated into two parts. The outer part is rounded and an extension of the lateral ridge; the inner part is acute and bears the typical specialized setae. The secondary process is less than one-third the length of the intermediate appendage. On the inferior appendage, the posterior promi-

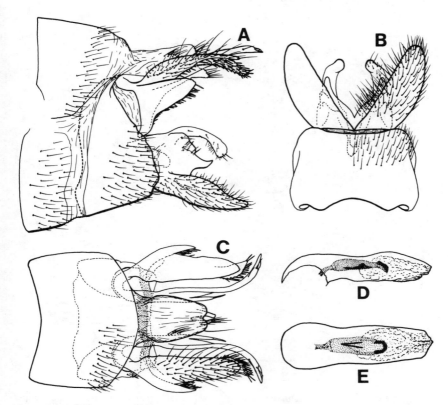

Figure 7-18. *Polycentropus domingensis* Banks.

nence of the vertical brace is well developed, but not so much as in *P. jeldesi*, and the dorsobasal process is slightly longer and thicker than in that species. The mesoventral processes on segment x are prominent. Male forewing length: 10.5 ($n = 1$).

Distribution. This species is known only from the types collected in the high Sierra Maestra (Pico Cuba, Massif Turquino) of southeastern Cuba, Oriente Province (Fig. 7-6).

Polycentropus domingensis Banks

Polycentropus domingensis Banks, 1941:399, Figs. 18–19, 23; Flint, 1967b: 6, Fig. 21, 1968b:80; Fischer, 1972:27; Flint, 1976:237, Figs. 7–8; Hamilton, 1986b:57–58, Fig. 3.3.

The preanal appendage of *P. domingensis* (Fig. 7-18) is uniquely different in form, being partially divided into three parts. The flaplike ventral part is without setae on its lateral surface and has specialized setae on its inner

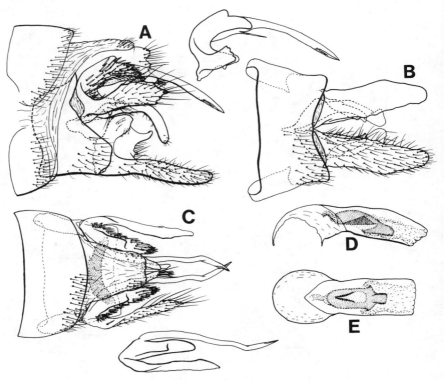

Figure 7-19. *Polycentropus vanderpooli* Flint, insets of left preanal and intermediate appendages without setae (lateral aspect—top, middle; dorsal aspect—bottom, middle).

surface. The two dorsal parts are closely aligned; the more basodorsal part bears normal, thin setae, and the more apicoventral part is acute and bears specialized setae. The secondary process of the intermediate appendage is similar to that in *P. turquino*. The vertical brace of the inferior appendage has a large, hooklike posterior prominence. Like *P. vanderpooli*, the phallobase lacks lateral flanges. Male forewing length: 10.8 mm ($n = 1$).

Distribution. This species is known only from types collected in the central and south-central Dominican Republic (Fig. 7-5).

Polycentropus vanderpooli Flint

Polycentropus vanderpooli Flint, 1976:237, Figs. 5–6; Hamilton, 1986b: 65–66, Fig. 3.10.

The preanal appendage of *P. vanderpooli* (Fig. 7-19) is divided into two prominent and distinct lobes: the outer lobe resembles the preanal appendage of less derived members of this species-group (*P. nigriceps* and *P. zaneta*), and the inner lobe bears numerous thick, specialized setae typical of the more derived members of the *nigriceps* group. The intermediate appendage has a short, ventrally positioned secondary process, which was not reported by Flint (1976). The vertical brace of the inferior appendage has a very large, hooklike posterior prominence similar to that in *P. domingensis.* The ventral portion of the inferior appendage is quite elongate, and the dorsobasal process is long and downcurved. As in *P. domingensis,* the phallobase lacks lateral flanges and is moderately depressed. Male forewing length: 6.2–7.7 mm (*n* = 7).

Distribution. This species is known only from the types collected in the mountains of the central and southeastern Dominican Republic (Fig. 7-6).

The *Insularis* Group

Polycentropus Insularis group, Flint, 1974:37; Hamilton, 1986b:122, 1987:145.

A diagnosis for this group has been provided by Flint (1980) and is supported by the abbreviated character analysis below. Females have been associated for only *P. insularis* (Flint 1968b) and *P. altmani* (Flint 1981). *Polycentropus insularis* is the only species for which the larva is known and biological data are available (Flint 1968b).

A cladistic analysis of the *insularis* group is presented in Hamilton (1987). This study focuses on the island species, on discovering the *insularis* group's sister taxon and its distribution, and on understanding this distribution in a historical context. Therefore, diagnoses for each species of this group will not be given as was done for species of the *nigriceps* group. However, each species of this group is readily separated from the others by details of the genitalic appendages and processes, much as in the *nigriceps* group. Synonymies and distribution summaries are provided below for each species.

Polycentropus altmani Yamamoto

Polycentropus altmani Yamamoto, 1967:130, Figs. 2A–C; Flint, 1968b:21 (transferred to *Polyplectropus*), 1981:15, Figs. 48–49 (as *Polycentropus*); Hamilton, 1986b:123–124, Fig. 6.32, 1987:145, Fig. 1.
Polycentropus macrostylus Flint, 1967a:8, Figs. 27–28, 1981:15, Figs. 48–49 (as jr. syn. of *P. altmani* Yamamoto).

Distribution. P. altmani (Fig. 7-20) is the most widespread species of the *insularis* group. I have seen records or specimens from Panama, Costa Rica, Nicaragua, Ecuador, and Venezuela (Fig. 7-7).

Polycentropus biappendiculatus Flint

Polycentropus biappendiculatus Flint, 1974:36, Figs. 60–62; Hamilton, 1986b:124–125, Fig. 6.33, 1987:145–146, Fig. 2.

Distribution. P. biappendiculatus is known from the holotype collected in central Surinam at Tafelberg (Table Moutain), apparently near a small

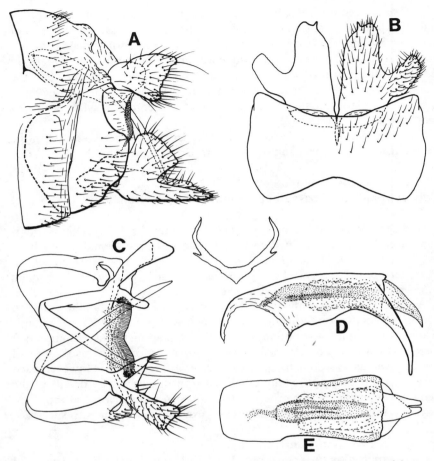

Figure 7-20. *Polycentropus altmani* Yamamoto, inset of subphallic support, caudal aspect. Terga VIII–X removed.

stream, and from a single male collected at the base of Cerro de la Neblina, Venezuela (Fig. 7-7).

Polycentropus cuspidatus Flint

Polycentropus cuspidatus Flint, 1980:149, Figs. 1–4; Hamilton, 1986b: 125–126, Fig. 6.34, 1987:146, Fig. 3.

Distribution. P. cuspidatus is known only from the holotype collected in central Ecuador near Puyo, Pastaza Province (Fig. 7-7).

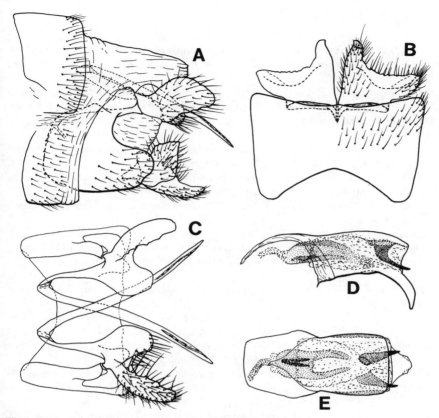

Figure 7-21. *Polycentropus insularis* Banks. Terga VIII–X removed.

Polycentropus insularis Banks

Polycentropus insularis Banks, 1938:302, Figs. 22, 24; Fischer, 1962:83; Flint, 1967b:6, Fig. 24, 1968b:24, Figs. 42–46; Hamilton, 1986b:127, Fig. 6.35, 1987:147, Fig. 5.

Distribution. *P. insularis* (Fig. 7-21) was described from Grand Etang, Grenada. Flint (1968b) has subsequently reported it from numerous locations on Dominica (Fig. 7-7). Specimens were collected from or near high elevation streams and lakes.

Polycentropus surinamensis Flint

Polycentropus surinamensis Flint, 1974:37, Figs. 63–66; Hamilton, 1986b: 128–129, Fig. 6.36, 1987:146, Fig. 4.

Distribution. *P. surinamensis* is known only from the type series collected in the Nassau Mountains of eastern Surinam (Fig. 7-7).

References

Banks, N. 1938. New West Indian neuropteroid insects. Rev. Entomol. 9:285–304.
———. 1941. New neuropteroid insects from the Antilles. Mem. Soc. Cubana Hist. Nat. 15:385–402.
Botosaneanu, L. 1978. Les Trichoptères de Cuba—faunistique, affinités, distribution, écologie. In M.I. Crichton, ed., Proceedings of the 2d International Symposium on Trichoptera, 1977, pp. 225–230. Dr. W. Junk Publ., The Hague. 359 pp.
———. 1980. Trichoptères adultes de Cuba collectés par les zoologistes cubains (Trichoptera). Mitt. Münch. Entomol. Ges. 69:91–116.
Coney, P.J. 1983. Plate tectonic constraints on the biogeography of Middle America and the Caribbean region. Ann. Missouri Bot. Gard. 69:432–443.
Croizat, L., G. Nelson, and D.E. Rosen. 1974. Centers of origin and related concepts. Syst. Zool. 23:265–287.
Curtis, J. 1835. Trichoptera. In British Entomology, vol. 12, pls. 530–577.
Davis, M.B. 1934. Habits of the Trichoptera. In C. Betten, ed., The Caddis Flies or Trichoptera of New York State, pp. 82–106. New York State Mus. Bull., no. 292. 576 pp.
Denning, D.G. 1947. New Trichoptera from Puerto Rico. Ann. Entomol. Soc. Amer. 40:656–661.
Donnelly, T.W. 1975. The geological evolution of the Caribbean and Gulf of Mexico—some critical problems and areas. In A.E.M. Nairn and F.G. Stehli, eds., The Ocean Basins and Margins, vol. 3: The Gulf of Mexico and the Caribbean, pp. 663–689. Plenum Press, New York. 706 pp.
———. 1985. Mesozoic and cenozoic plate evolution of the Caribbean Region. In F.G. Stehli and S.D. Webb, eds., The Great American Biotic Interchange, pp. 89–121. Plenum Press, New York. 532 pp.
Erwin, T.L. 1983. Tropical forest canopies: The last biotic frontier. Bull. Entomol. Soc. Amer. 29(1):14–19.
Fischer, F.C.J. 1962. Polycentropodidae, Psychomyiidae. Trichopterorum Catalogus, vol. III. Nederl. ent. Ver., Amsterdam. 236 pp.
———. 1972. Supplement to vols. III and IV. Trichopterorum Catalogus, vol. XIII. Nederl. ent. Ver., Amsterdam. 172 pp.
Flint, O.S., Jr. 1964. The caddisflies (Trichoptera) of Puerto Rico. Univ. Puerto Rico Agric. Exp. Sta. Tech. Pap. 40:1–80.
———. 1967a. Studies of Neotropical caddis flies, IV: New species from Mexico and Central America. Proc. U.S. Natl. Mus., 123(3608):1–24.
———. 1967b. Studies of Neotropical caddisflies, V: Types of species described by Banks and Hagen. Proc. U.S. Natl. Mus., 123(3619):1–37.
———. 1968a. The caddisflies of Jamaica (Trichoptera). Bull. Inst. Jamaica, Sci. Ser. 19:1–68.
———. 1968b. Bredin–Archbold–Smithsonian Biological Survey of Dominica, 9. The Trichoptera (caddisflies) of the Lesser Antilles. Proc. U.S. Natl. Mus. 125 (3665):1–86.
———. 1974. Studies of Neotropical caddisflies, XV: The Trichoptera of Surinam. Studies on the Fauna of Suriname and other Guyanas 14:1–151. 4 pls.

———. 1976. The Greater Antilles species of *Polycentropus* (Trichoptera: Polycentropodidae). Proc. Biol. Soc. Washington 89(17):233–246.

———. 1978. Probable origins of the West Indian Trichoptera and Odonata faunas. *In* M.I. Crichton, ed., Proceedings of the 2d International Symposium on Trichoptera, 1977, pp. 215–228. Dr. W. Junk Publ., The Hague. 359 pp.

———. 1980. Studies of Neotropical caddisflies, XXIX: The genus *Polycentropus* (Trichoptera: Psychomyiidae). J. Washington Acad. Sci. 70(4):148–160.

———. 1981. Studies of Neotropical caddisflies, XXVIII: The Trichoptera of the Rio Limon Basin, Venezuela. Smithsonian Contrib. Zool., no. 330. 61 pp.

Hamilton, S.W. 1986a. A new species of Cuban *Polycentropus* (Trichoptera: Polycentropodidae). Proc. Entomol. Soc. Washington 88:731–733.

———. 1986b. Systematics and biogeography of New World *Polycentropus sensu stricto* (Trichoptera: Polycentropodidae). Ph.D. diss. Clemson University. 257 pp.

———. 1987. Phylogeny of the *Polycentropus insularis* species-group (Trichoptera: Polycentropodidae). *In* M. Bournaud and H. Tachet, eds., Proceedings of the 5th International Symposium on Trichoptera, pp. 145–148. Dr. W. Junk Publ., Dordrecht. 397 pp.

Hedges, S.B. 1982. Caribbean biogeography: Implications of recent plate tectonic studies. Syst. Zool. 31:518–522.

Humphries, C.J. 1982. Vicariance biogeography in Mesoamerica. Ann. Missouri Bot. Gard. 69:444–463.

Kingsolver, J.M. 1964. New species of Trichoptera from Cuba. Proc. Entomol. Soc. Washington 66:257–259.

Lepneva, S.G. 1964. Larvae and pupae of Annulipalpia. Fauna of the U.S.S.R., Trichoptera 2(1):1–560. (English translation, U.S. Dept. Commerce, Springfield, Va., 1970.)

McCafferty, W.P. 1981. Aquatic Entomology. Science Books International, Boston, Mass. 448 pp.

Malfait, B.T., and M.G. Dinkleman. 1972. Circum-Caribbean tectonic and igneous activity and the evolution of the Caribbean Plate. Geol. Soc. Amer. Bull. 83:251–272.

Nelson, G., and N.I. Platnick. 1981. Systematics and Biogeography: Cladistics and Vicariance. Columbia University Press, New York. 567 pp.

Nelson, G., and D.E. Rosen, eds. 1981. Vicariance Biogeography: A Critique. Columbia University Press, New York. 593 pp.

Pindell, J., and J.F. Dewey. 1982. Permo-Triassic reconstruction of western Pangea and the evolution of the Gulf of Mexico/Caribbean region. Tectonics 1:179–211.

Platnick, N.I. 1976. Drifting spiders or continents? Vicariance biogeography of the spider subfamily Laroniinae (Aranea: Gnaphosidae). Syst. Zool. 25:101–109.

Platnick, N.I., and G. Nelson. 1978. A method of analysis for historical biogeography. Syst. Zool. 27:1–16.

Rosen, D.E. 1974. Phylogeny and zoogeography of salmoniform fishes and relationships of *Lepidogalaxias salamandroides*. Bull. Amer. Mus. Nat. Hist. 153:265–326.

———. 1975. A vicariance model of Caribbean biogeography. Syst. Zool. 24:431–464.

————. 1978. Vicariant patterns and historical explanation in biogeography. Syst. Zool. 27:159–188.

————. 1979. Fishes from the uplands and intermontane basins of Guatemala: Revisionary studies and comparative geography. Bull. Amer. Mus. Nat. Hist. 162:267–376.

————. 1985. Geological hierarchies and biogeographic congruence in the Caribbean. Ann. Missouri Bot. Gard. 72:636–659.

Ross, H.H. 1944. The caddisflies, or Trichoptera, of Illinois. Bull. Illinois Nat. Hist. Surv. 23:1–326.

————. 1967. The evolution and past dispersal of the Trichoptera. Ann. Rev. Entomol. 12:169–206.

Savage, J.M. 1982. The enigma of the Central American herpetofauna: Dispersal or vicariance? Ann. Missouri Bot. Gard. 69:464–547.

Schmid, F. 1968. Le genre *Poecilopsyche* n. gen. Ann. Soc. Entomol. Quebec 13:3–31.

————. 1980. Les Insectes et Arachnides du Canada, partie 7: Genera des Trichoptères du Canada et des États Adjacents. Agric. Canada, Ottawa. 296 pp.

————. 1984. Essai d'evaluation de la faune mondiale des Trichoptères. *In* J.C. Morse, ed., Proceedings of the 4th International Symposium on Trichoptera. Dr. W. Junk Publ., The Hague. 486 pp.

Sparks, A.N., R.D. Jackson, J.E. Carpenter, and R.A. Muller. 1986. Insects captured in light traps in the Gulf of Mexico. Ann. Entomol. Soc. Amer. 79:132–139.

Unzicker, J.D., V.H. Resh, and J.C. Morse. 1982. Trichoptera. *In* A.R. Brigham, W.U. Brigham, and A. Gnilka, eds., Aquatic Insects and Oligochaetes of North and South Carolina, pp. 9.1–9.138. Midwest Aquatic Enterprises, Mahomet, Ill. 837 pp.

Wiggins, G.B. 1977. Larvae of the North American Caddisfly Genera (Trichoptera). University of Toronto Press, Toronto. 401 pp.

Wiley, E.O. 1981. Phylogenetics: The Theory and Practice of Phylogenetic Systematics. Wiley Interscience Publ., New York. 439 pp.

Wolcott, G.N. 1950. The insects of Puerto Rico. J. Agric. Univ. Puerto Rico 32:1–224.

Yamamoto, T. 1967. New species of the caddisfly genus *Polycentropus* from Central America (Trichoptera: Polycentropodidae). J. Kansas Entomol. Soc. 40:127–132.

8 · Relicts in the Drosophilidae (Diptera)

David A. Grimaldi

So far as I know, almost all continental drosophilids have ranges that cover an area at least an order of magnitude greater than those of insular species. Also, closely related mainland species are sympatric more often than not. Assuming an allopatric model of speciation, these two facts must mean that drosophilids have considerable ability to disperse and colonize, which may explain why the endemicity of drosophilids in the Caribbean region is quite low compared with many other insects, such as some carabids (see other papers in this volume). This hypothesis will be tested here, as applied to Caribbean Drosophilidae.

Fifty-eight species of Drosophilidae are endemic to the Antilles, 11 of which have yet to be described (App. 8.1). At least 12 genera and subgenera are represented in the region (68 world genera have been described for about 2500 species, and some await classification). I can discuss in detail only three groups here: *Mayagueza*, the *Drosophila repleta* species group, and *Zygothrica*. *Mayagueza* is probably the most plesiomorphic and *Zygothrica* the most recently derived of the three taxa. The decision to compare these three taxa was based on several criteria: for *Mayagueza*, because it is the only endemic genus in the Caribbean for the Drosophilidae; for the *D. repleta* species-group, because extensive research has already been done; and for *Zygothrica*, because it is a taxon I am currently studying.

Throughout the text there will be references to the age or relative age of a group. Given three species or taxa of other rank (A, B, and C), one set of relationships may be that A is the sister group to B + C. In this event, the cladogenesis between A and B + C is certainly no younger than, and may have preceded the event that gave rise to, B and C. Thus the lineage to which A belongs (and perhaps A itself) is either as old or older than either lineage B or C. These definitions are important to understand my use of the word "relict."

A relict (or paleoendemic, or anachronism) is a narrowly endemic, living member of a relatively old group. "Old" is taxonomically subjective: it depends on genealogy and where the presumed relict lineage arises in a phylogenetic tree. Implied in the endemicity of a relict lineage is extinction: the relict is a remnant surrounded by extinction, it is part of a waning lineage. In this paper I try to be explicit about the relative age of some Carribbean drosophilids. It is often difficult to ascertain whether any one species is itself a relict because of the difficulty in identifying the extinct ancestors of an endemic taxon or the extinction of the relict itself in many parts of its former range (see, e.g., many examples in Eldredge and Stanley 1984). Therefore, my use of "relict" is qualified to mean a relict *lineage* to which the endemic belongs. As I discuss later, a study of Dominican and Chiapas amber fossils (Grimaldi 1987a) has thus far provided little data to corroborate the role of inferred extinctions in present endemism patterns of Caribbean drosophilids.

Results

Mayagueza

Mayagueza, a monotypic genus, was described for a species presently known only from Mayaguez, Puerto Rico. *M. argentifera* was apparently given generic rank because of several features that it shares with various genera of lower Drosophilidae (Wheeler 1960). I will present explicit evidence for *Mayagueza* genealogy.

Mayagueza belongs to the subfamily Steganinae, whose members usually possess several more primitive traits than those in the other subfamily, the Drosophilinae. There is strong indication that the Steganinae are paraphyletic, namely, they are defined either by the possession of traits that arose well before the Drosophilidae (e.g., strong prescutellar setae and several other chaetotaxal characteristics, and the absence of "teeth" on the "egg guide" [= sternite VIII]), or by several clearly derived traits that occur sporadically in the subfamily (e.g., "pegs" on the distal costal wing vein segment). The *Pseudiastata* genus-group, to which *Mayagueza* belongs, will be designated here in the steganine confusion.

To examine the relationships of *Mayagueza*, it was necessary to study representative steganines, particularly those with pubescent aristae, such as *M. argentifera* (Fig. 8-2C–E). Màca (1980b) believed this state of the arista to be primitive, but when comparing it with several drosophilid relatives (using the taxa mentioned below), it seems obvious that the trait is a derived one for the Drosophilidae. Adults belonging to 15 and the larvae of 6 of the following genera/subgenera were examined (* indicates immature material):

Acletoxenus indicus,* *Acletoxenus* sp.* (Sumatra), *Amiota* (*Amiota*) *humeralis*, *Amiota* (*Sinophthalmus*) *picta*, *Amiota* (*S.*) *polychaeta*, *Apenthecia crassiseta*, *Cacoxenus* (*Gitonides*) *perspicax,** *Cacoxenus* (*Paracacoxenus*) *guttatus,** *Gitona bivisualis*, *G. brasiliensis*, *Leucophenga varia*, *Leucophenga* spp. (4, Neotropical), *Mayagueza argentifera*, *Pseudiastata pseudococcivora,** *P. vorax,** *Pseudiastata* spp. (2, Mexico, Trinidad), *Rhinoleucophenga obesa,** *R. pallida*, *Rhinoleucophenga* spp. (2, Trinidad, Panama), *Stegana coleoptrata*, *Stegana* sp. near *tarsalis*, *Stegana* spp. (4, Neotropical).

Many of the above taxa do not have pubescent aristae, but they were included in the study to better judge steganine variation. In addition, the following papers served as references for some taxa that I did not examine, either as adults or as larvae/puparia: McAlpine (1968), Màca (1977, 1980a, 1980b), Okada (1968), and Wheeler (1960). Gonçalves (1939) provided a good description of the habits and morphology of *Pseudiastata brasiliensis* larvae. For outgroup taxa, which were used to decide on the primitive state for characters, the following were used: *Diastata repleta*, *D. eluta* (Diastatidae); *Camilla glabra* (Camillidae); *Notiphila teres* (Ephydridae); *Curtonotum helvum* (Curtonotidae). The published information on ephydrid immatures helped to polarize the states of immature characters at the family level for the Drosophilidae (diastatid and camillid immatures are unknown).

Figures 8-1 and 8-2 and Appendix 8.2 present some of the taxonomic characters. Based on the cladogram (Fig. 8-3), I concur with Wheeler (1960) that at least *Pseudiastata* and *Acletoxenus* are close relatives of *Mayagueza*, but *Cacoxenus* is not. *Mayagueza* is also related more closely to *Acletoxenus* than to *Pseudiastata*. Before turning to the biogeography of the flies, I will briefly discuss two interesting implications of the cladogram. First, there are at least three major clades among the Steganinae: (1) one to which belong the genera *Stegana* and *Leucophenga* which are phenetically quite distinct; (2) a large one that separates the subgenera of the large genus *Amiota*; and (3) the *Pseudiastata* genus group. More taxa and characters need to be examined before formal categories are assigned to any of the clades. Second, among the traits that appear to be homoplasious, the most interesting one is predation by the larvae. A very readable and thorough review of the habit is given in Ashburner (1981). Obligate predation by larvae has arisen four or perhaps five times among the taxa in Figure 8-3, and several more times (e.g., *Cladochaeta*, *Titanochaeta*) in all the Drosophilidae (see Figs. 8-9A, B for two of the hosts). Several African *Leucophenga* larvae feed on cercopids, *Cacoxenus perspicax* feeds on coccids, *Rhinoleucophenga obesa* is a predator of *Aclerda* scales, and *Gitona brasiliensis* predates mostly *Orthezia* among some other scales. The existence of different prey taxa confirms that predation at these levels has appeared independently. The habits of many *Stegana* and *Amiota*, however, are unknown. True to the predatory habits of the

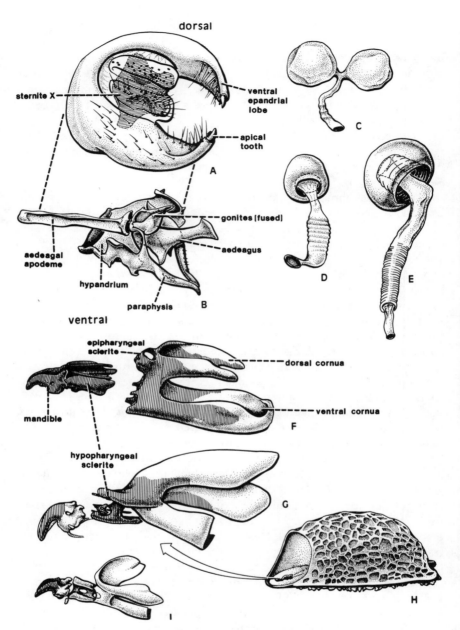

Figure 8-1. Some taxonomic characters used in the cladogram, Fig. 8-3. A, B: Male genitalia of *Mayagueza argentifera*. A, epandrium; B, genitalia. C–E: spermathecal capsules (to same scale as A, B). C, *M. argentifera* (fusion of the pair is a very unusual trait in the Drosophilidae); D, *Acletoxenus* sp. (India); E, *Amiota* (*Erima*) *crassiseta*. F, G, I: larval instar III cephalopharyngeal skeletons (extracted from puparia). F, *Cacoxenus* (*Gitonides*) *perspicax*; G, *Pseudiastata pseudococcivora*; I, *Acletoxenus* sp. (India). H: puparium of *Pseudiastata pseudococcivora*.

186

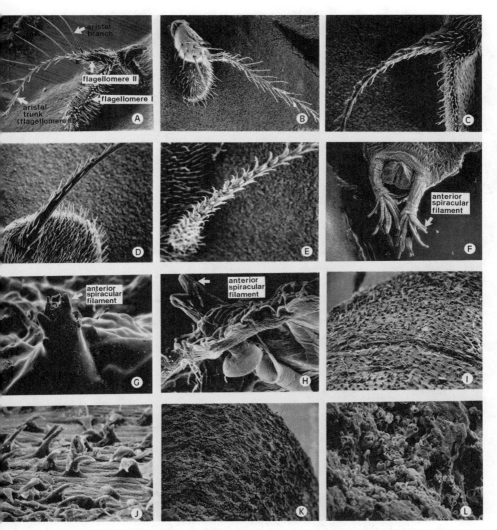

Figure 8-2. Scanning electron micrographs of some structures used for steganine phylogenetic reconstruction. A–E: antennae or portions thereof. A, *Camilla glabra* (Camillidae), 170×; B, *Diastata eluta* (Diastatidae), 74× (short plumose); C, arista (flagellomeres II & III) of *Acletoxenus indicus* (Drosophilidae), 207×; D, basal portion of arista of *Amiota* (*Sinophthalmus*) *picta* (Drosophilidae), 218×; E, detail of arista of *Mayagueza argentifera* (Drosophilidae), 183×; F, anterior spiracles of *Drosophila* (*Siphlodora*) *busckii*, 56×; G, detail of anterior spiracle of *Pseudiastata nebulosa* pupa, 285× (note stubby filaments); H, anterior spiracle (also showing internal attachement of trachea and felt chamber) of *Acletoxenus* sp. (India), 659×; I, dorsal puparial integument, *Cacoxenus* (*Paracacoxenus*) *guttatus*, 97×; J, detail of spicules, showing bifid structure, *Cacoxenus guttatus*, 500×; K, dorsolateral puparial surface, *Pseudiastata nebulosa*, 41×; L, detail of view in K, showing poroid surface and curled (waxy?) exudate, 422×.

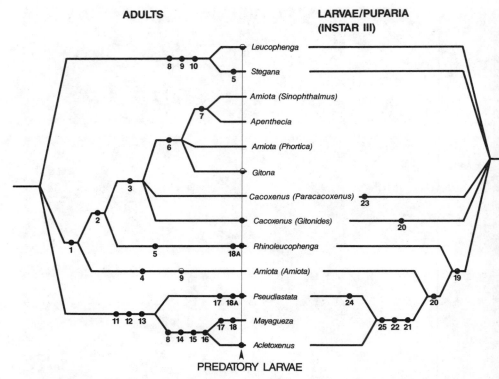

Figure 8-3. Hypothesis of phylogenetic relationships among some genera and subgenera of steganine Drosophilidae, which is based on immature characters (cladogram at right) and characters of adult morphology (left). The appearance of predatory larvae has occurred several times independently, which is indicated (bicolored circles indicate that the predatory habit is found among some of the members of that taxon).

lower Drosophilidae are the tastes of *Pseudiastata* and *Acletoxenus*: the former feeds on scales, the latter on white flies (Aleurodidae: Homoptera). It takes very little imagination to predict that when *Mayagueza argentifera* larvae are found, they will be predatory on sessile Homoptera and will share some traits of the cephalopharyngeal skeleton that presently are restricted to the *Pseudiastata* genus group.

Acletoxenus is a widespread genus, due primarily to the distribution of *A. formosus*, which occurs in Europe, the Middle East, northern Africa, southeast Asia, and northern Australia. The remaining four species are found in tropical India, southeast Asia, and Australia (Fig. 8-4), and one of these is an undescribed species from Sumatra. Together, *Mayagueza* + *Acletoxenus* is the sister group of *Pseudiastata*, which is a predominantly Neotropical group of at least ten species; *P. nebulosa* is in the eastern and south-central United States, and I have seen specimens of two undescribed species from Mexico

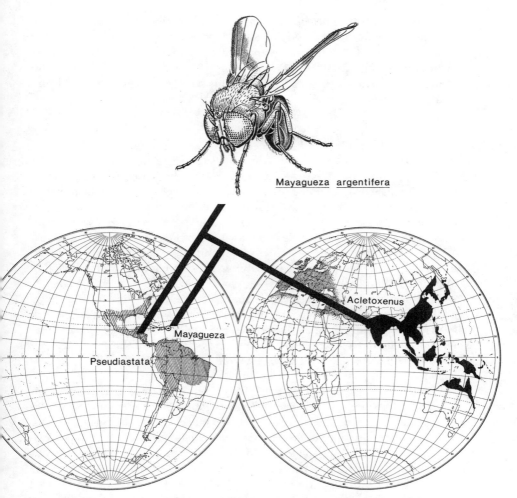

Mayagueza argentifera

Acletoxenus

Mayagueza

Pseudiastata

Figure 8-4. Distribution of the *Pseudiastata* genus group. *Pseudiastata* is hatched, *Mayagueza* occurs on Puerto Rico, and *Acletoxenus* spp. are in black (the range of *Acletoxenus formosus*, where it extends farther than its congeners, is stippled).

and Trinidad. Two hypotheses may account for the disjunct distribution of *Pseudiastata* genus group flies. One of them is more probable because it is supported by geologic evidence, but both presume that some African and Malagasy extinction of *Acletoxenus* sp(p). or a direct ancestor thereof has occurred.

On the one hand, dispersal of an *Acletoxenus-Mayagueza* ancestor from Puerto Rico to Africa may have left behind what is now *Mayagueza*, while also establishing *Acletoxenus*. But why would the Neotropical mainland not be colonized several times before central (tropical) Africa? I believe that the ancestor of the *Acletoxenus-Mayagueza* clade evolved around the time of the split between the Greater Antilles and South America, an event that simultaneous with or later than the African–South American divergence about 80 Ma. Both genera may be recent additions to what are actually old and vicariant lineages.

The *Drosophila repleta* species group

The *Drosophila repleta* group is one of 22 species groups recognized in the subgenus *Drosophila*. It is a well-defined group of higher Drosophilidae, and it is distinguished among other groups in the subgenus in part by the presence of a spotted integument or elaborate maculations. With 76 described morphospecies (Vilela 1983) and 3 species yet to be described (M. Wasserman, pers. comm. 1986), this is the second largest clade endemic to the New World. Most species are found in dry habitats, where the species breed mostly in cactus necroses. The remaining species are mostly forest-edge flies that breed in flowers and fruits.

The species group has been the subject of extensive study, primarily by William Heed and his students and by Marvin Wasserman, whose work spans three decades. The evolution and especially the ecology of these flies were presented in a recent book (Barker and Starmer 1982), from which my discussion has been distilled; in a review by Wasserman (1982), and in Vilela's revision (1983).

Wasserman split the species group into nine subgroups based on cytological characters (Fig. 8-5). The characters are found in the metaphase complements and the polytene chromosomes in the larval salivary glands. The latter are huge, presumably because of immense metabolic and transcriptional activity. Arranged along the length of each polytene chromosome are bands (chromomeres) of varying widths that also occur in a sequence characteristic of one arm or a portion thereof. If inversions or their portions are shared among two or more species, that chromosomal segment is considered a synapomorphy and can be used to infer genealogy. There is little doubt that inversions are truly unique character states (one idea is that smaller inversions may actually be transposons), but deciding which inversional sequence is primitive can be difficult. If it is impossible to discover an inversion in an outgroup taxon, morphological or other criteria can be used to polarize the trend in inversional states.

Two predominant biogeographic patterns involve endemic species on the Antilles. In one, old endemics are restricted to Jamaica, Puerto Rico, and Hispaniola. In the other the insular endemics have some close relatives that are widespread, usually in the southern United States and Central America, or in Central America plus northern South America. The following is a brief

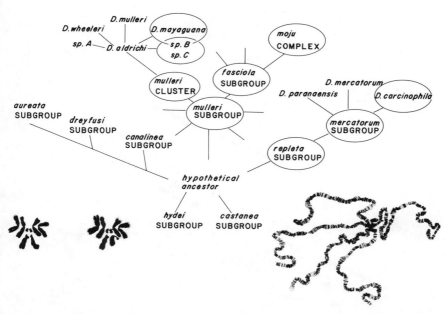

Figure 8-5. Relationships among groups in the *Drosophila repleta* species group (redrawn from Wasserman 1982). This phylogenetic tree has been constructed on polytene and metaphase chromosome characteristics. Taxa that are circled are Caribbean or have Caribbean representatives.

discussion of four endemic *repleta* lineages, three of which appear to be relatively old; the other has widespread relatives on the mainland, but the relationships are obscure.

Drosophila peninsularis, which is found in southern Florida, Puerto Rico, and Cuba (Fig. 8-6A), is one branch of a trichotomy that also involves the *repleta* and *fulvimaculata* complexes of species. This conclusion is based on cytological grounds, since Vilela (1983) places it in the *mercatorum* subgroup because of morphological resemblance. That the *repleta* subgroup probably originated before the *mercatorum* subgroup (which has one species endemic to the Caribbean) suggests *D. peninsularis* to be old indeed. Members of the *repleta* subgroup are widespread in deserts or scrub habitats abundant with cacti, such as in the southwestern United States, northern Mexico, Brazilian *caatingas*, and portions of Mato Grosso.

D. paraguttata, which occurs on Jamaica, is basal to the *moju* species complex. The other members of this complex are *D. mojuoides* (Trinidad) and *D. moju* (Costa Rica to Bolivia and southern Brazil) (Fig. 8-6B). According to Wasserman, the latter species should be separated into a Central American and a South American population. Here again is an insular endemic, *D. paraguttata*, that appears to be relatively old.

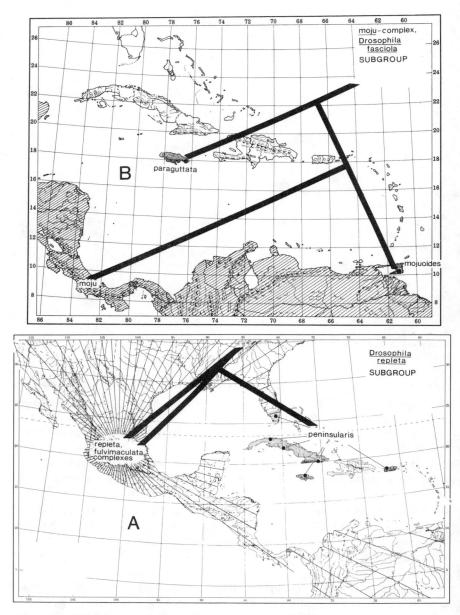

Figure 8-6. Distributions and relationships of some *Drosophila repleta* species group lineages (relationships from Wasserman 1982).

In the *mercatorum* subgroup are three species: *D. mercatorum, D. para-naensis*, and *D. carcinophila* (which occurs on the Bahamas, Cuba, Dominica, Grand Cayman, Jamaica, Montserrat, and Puerto Rico). *D. mercatorum* is very opportunistic in its use of hosts, facultatively parthenogenetic, and ranges over most of the southeastern United States and Central and South America; recently it has even colonized Hawaii and Africa. Curiously, there are no records or specimens of this species from the Antilles. Everywhere, its breeding sites consist mostly of decaying fruits and vegetables. The habits of *D. paranaensis* have yet to be discovered. *D. carcinophila* has specialized larvae that feed on the exudates in nephric grooves of the land crab, *Gecarcinus ruricola* (Carson 1967; Fig. 8-9E, F). Because it does not share two polymorphic inversions found in *D. paranaensis* and *D. mercatorum*, this fly seems to be the most primitive species of the *mercatorum* subgroup. The habit of crab commensalism has actually evolved three times in the Drosophilidae; the other two instances involve *Lissocephala powelli* (on Christmas Island) and *D. endobranchia* (Carson 1974). The latter species is also a Caribbean endemic (Cayman Island; perhaps Cuba, but this record is based only on larval specimens). The relationships of *D. endobranchia* are not entirely clear because it does not fit into any of the standard species-groups. Despite their highly derived tastes, the Caribbean "crab flies" appear to be recent members of nonterminal lineages.

Other insular endemics in the species group belong to the *mulleri* species cluster, which comprises *D. mayaguana* (widespread), and two new species, n.sp. 1 (Haiti, Navassa Island), and n.sp. 2 (Hispaniola, Jamaica). Wassermand and others believe that these three species and three others distributed from Nebraska to Venezuela are derived from *D. aldrichi* or a close relative thereof (Fig. 8-5). *D. aldrichi* has a very expansive range that extends from Texas to Colombia. Finally, there is an undescribed species from Jamaica that breeds well in the laboratory and has been examined morphologically and karyotypically (W. Heed, pers. comm. 1986). Like *D. endobranchia*, which will be discussed later, it shares derived features of two species groups, in this case the *repleta* and *nannoptera*.

Zygothrica

Zygothrica is the most speciose lineage of Drosophilidae in the New World (discounted here are the numerous species of *Drosophila*; this genus appears paraphyletic [see Throckmorton 1975]). Sixty-four species names presently exist, and a recent study gives descriptions of 48 additional species in a portion of the genus (Grimaldi 1987b). I currently estimate a total of about 200 species of *Zygothrica*. Seven of the species belong to an Indo-pacific clade (Samoa, east to Thailand on the Malay peninsula, and in northern Australia); members of the other clades occur from southern Mexico to Bolivia and southern Brazil. Despite their diversity on the Neotropical mainland, only 14 species are found on the Caribbean islands and 9 are endemic there (Table 8-1). Species epithets given below in quotations indicate taxa to be described in my upcoming monograph on the genus.

Before I discuss the Neotropical *Zygothrica*, I will point out a distribution found in one clade of the group that closely reflects the situation in the *Pseudiastata* genus group. Two New World species, *Z. bilineata* (which is Amazonian and circumcaribbean) and *Z.* "flavifrons" (presently known only from Trinidad), are the closest relatives of the Indopacific, *samoaensis*-group species. As with the *Pseudiastata* group, no African relatives exist. It is very difficult to say if the same geologic events have affected the distributions of the two unrelated groups.

All species of *Zygothrica* are known to aggregate at imbricate polypores and Tricholomataceae, particularly white fleshy ones (Fig. 8-9C). Some primitive species breed in the fungi, but most breed in fleshy bracts of flowers, principally in the Zingiberales (Fig. 8-9D), and simply retain the plesiomorphic habit of using fungi to rendezvous. During dry seasons, when fungus blooms are scarce, the aggregations of *Zygothrica* are very dense and can approach a "standing room only" situation.

My phylogenetic hypothesis on the genus is presented in Figure 8-7. It is based on the morphology of 53 species and the reproductive behavior repertoires of 7 of them. The majority of species fit well into particular clades, but the relationships among clades are not clear. I do not believe this polychotomous lineage exhibits paraphyly: the synapomorphies distinguishing the genus are distinctive for drosophilids, and behavior corroborates the morphological conclusions. Most *Zygothrica* are lowland tropical, moist/rain forest dwellers (<500 m), but a few are restricted to high altitudes (>1500 m). None of the Caribbean endemics is a highland species: only *Z. microstoma*, which is circumcaribbean, is a highland species in the Greater Antilles (Table 8-3).

Table 8-1. Nonendemic Caribbean *Zygothrica*

Z. microstoma Duda, 1927 (1.1, microstoma group): Puerto Rico, Jamaica, southern Mexico to trans-Andean Ecuador and northern Peru

Z. circumveha Grimaldi, 1987 (1.2.1, *dispar* group): Haiti, Jamaica, southern Mexico to northern Colombia

Z. vitticlara Burla, 1956 (1.2.2, unnamed group): Dominica, St. Lucia, Brasil (São Paulo)

Z. bilineata (Williston) 1896 (2.1, *bilineata-samoaensis* group): Cuba, Dominica, Guadeloupe, Haiti, Jamaica, Martinique, Puerto Rico, St. Vincent, from Panama (Chiriqui) to Rio de Janeiro, Brasil

Z. atriangula Duda, 1927 (4.2.2, *atriangula* group): Dominica, Jamaica, southern Vera Cruz (Mexico) to São Paulo, Brasil, and Cocos Is. (Costa Rica)

NOTES: Major lineage to which each species belongs is in parentheses. Information based on cladogram in Figure 8-7.

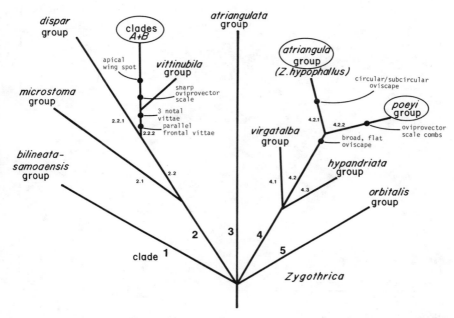

Figure 8-7. Hypothesis of phylogenetic relationships among species groups of the genus *Zygo-thrica*. Clade 1 is Neotropical and Indopacific in distribution, 2–5 are entirely Neotropical. Groups that are circled have Antillean representatives.

Little is known about three endemic *Zygothrica*—*Z.* "palpivanna," *Z. vitrea*, and *Z.* "hypophallus." Specific placement of the first two of these undescribed species, which are found on Jamaica, must await the discovery of females. *Z.* "hypophallus" is a modified member of a terminal clade that includes some very widespread species (e.g., *Z. atriangula*; Table 8-1), and is known only from St. Vincent.

Two major groups of Caribbean Drosophilidae exist. First is the *poeyi* clade, which includes *Z. poeyi*, *Z. semistriata* Wheeler, and *Z.* "para-semistriata" (note that this is not the "poeyi" of Burla 1956, which is actually *Z. laevifrons* Duda, a widespread Amazonian species). *Z. poeyi* is the sister group to the pair of other species, and it occurs from Cuba to Puerto Rico. Collections on Jamaica, Puerto Rico, and the Dominican Republic have been good, so *Z.* "parasemistriata" is probably restricted to the Haitian peninsula, assuming that the rampant deforestation in this area has not taken its toll. *Z. semistriata* is known only from southern Mexico on the Gulf Coast. The endemism pattern of the *poeyi* species group is another instance where a Greater Antillean species is the sister group to a Central American clade or species (Fig. 8-8A).

The distribution of the second *Zygothrica* clade is shown in Figure 8-8B.

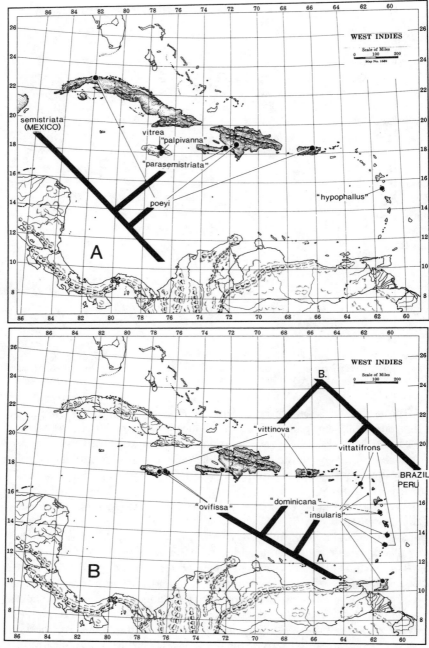

Figure 8-8. *Zygothrica* distributions and relationships. A, assorted species (*Z.* "*vitrea*" and *Z.* "*palpivanna*" are as yet unplaced; *Z.* "*hypophallus*" belongs to the *atriangula* species group) and species of the *poeyi* species group. B, the two lineages of *Zygothrica* clade 2.2.2. that occur in the Caribbean region.

196

The species in this group, like numerous superficially similar ones, are attractive, striped forms with apical wing spots (Fig. 8-10), and they form two main lineages (shown in Fig. 8-8B as A and B). The monophyly of A + B is supported by a trait unique in Diptera: the females possess one to three terminal, large, ramphate scales at the apex of the eversible ovipositor membrane (the oviprovector). The scales embed themselves in the host plant tissue during oviposition, probably to anchor the female.

The distribution of clades A and B can best be described as countercurrent. Outgroup comparison places the primitive state for the distribution of this group as Amazonian. In clade A, the first divergence event is between the pair Z. "insularis" + n.sp. from Peru (700 m altitude) (not shown in Fig. 8-8B) and the pair Z. "dominicana" (on Dominica) + Z. "ovifissa" (on Jamaica, Haitian peninsula). Even if this distribution is not a progression up the Lesser Antilles, there is little doubt that Z. "ovifissa" is a relatively recent addition to the Greater Antilles.

Clade B has a distribution with an opposite direction. Z. "vittinova" (on Jamaica, Puerto Rico, and perhaps Hispaniola) is the most plesiomorphic member of clade B, and it is the sister group to the following set of species: Z. vittatifrons (St. Kitts, St. Vincent, and probably the intervening islands) + n. sp. from Brazil (São Paulo State) + n.sp. from Peru (mid-altitude). There is a problem with the interpretation that B is spreading southeastward, for it suggests that the lineage B evolved in the Greater Antilles, which A has only recently colonized. Are A + B paraphyletic? Because there appears to be no reason to revise the morphological decisions, a remaining explanation for the countercurrent distributions of clades A and B is extinction. The ancestor to A + B might have been widespread on both the mainland and the islands, given rise to each lineage in different areas, and gone extinct, since no extant and plesiomorphic fly has such a distribution.

Among the three taxa discussed in this paper, Zygothrica appears to have the greatest number of recent additions to the Antilles, composed of Z. "hypophallus" and clades A + B. The other lineages that have endemic species also have mainland relatives that appear to be more recently derived. Even though it should not be standard practice in historical biogeography to discuss which groups are not present in an area, especially where extinction is of primary concern, the absence of species belonging to the dispar species group (clade 2.2.1 in Fig. 8-7) is a glaring attribute of the West Indian drosophilid fauna. Except for the primitive and circumcaribbean species Z. "circumveha," no others in this group are found on the islands. And this is despite the fact that some dispar group members are among the most common Zygothrica: Z. prodispar, for example, has the most expansive distribution in Zygothrica (Vera Cruz, and Mexico to Bolivia). Their absence from the islands is perplexing and may be evidence that dispersal in Caribbean Drosophilidae was not important.

Other Caribbean Drosophilidae

Three other drosophilid groups have Caribbean endemics that are cladistically basal to mainland groups. First is *D. insularis*, a member of the *willistoni* species-group. It is endemic to St. Kitts and St. Lucia. Except for *D. neoalagitans*, the other species in the group are widespread and abundant rain forest inhabitants in Central and South America. Dobzhansky (1957) mentioned that, based on morphological and chromosomal grounds, the relationships of *D. insularis* are obscure, for it possesses few of the derived traits in the group. Second, there is *D. endobranchia*, which, as noted earlier, is a perplexing species: it possesses traits of the *D. quinaria* and *D. virilis* species groups, both of which are widespread Holarctic and entirely continental groups. Third is *Paramycodrosophila nephelea* on Jamaica. There are ten other *Paramycodrosophila* species: one in the southeastern United States, seven in the Indopacific and in northern Australia, and two circum-Caribbean species that include Hispaniola and Puerto Rico in the Antillean portion of their distributions. The *Paramycodrosophila* distribution is very similar to that of the *Pseudiastata* genus group and to some *Zygothrica* that I have already mentioned.

An addition to Cuba that is probably recent is *Chymomyza microdiopsis*, and one recent to Jamaica is *C. jamaicensis* (Grimaldi 1986). Each of these two species belongs to the two main lineages in the *C. aldrichii* species group. They represent terminal clades, and the species group itself is cladistically terminal compared to the other groups in the genus (Okada 1976). The closest relative of *C. microdiopsis* is *C. exophthalma*, a species ranging from Panama to Peru. *C. guyanensis* + *diatropa* is the sister group to *C. jamaicensis*, and the range of this group is from Costa Rica to Guyana, and the southern tip of Florida (*C. diatropa*). The sister group of the *C. aldrichii* species group is the *procnemis* species group, which is most speciose in the Congo Basin and has several members in southeast Asia (*C. procnemis* is the only Nearctic representative). Since the *aldrichii* species group is entirely New World, it is certainly no older than the split between South America and Africa about 80 Ma and may be much more recent than that.

Discussion

With the exception of *Zygothrica* clade "A," *Z.* "hypophallus," and the two *Chymomyza*, the Antillean Drosophilidae probably appeared before the divergence in their mainland relatives. This is the pattern first elucidated by Rosen (1975:456, Fig. 20), whereby taxa from the Greater Antilles represent the sister groups to taxa from Central America + northern South America. It is apparent from the discussion of the three drosophilid taxa that these

Figure 8-9. Hosts and prey of some drosophilid taxa that have Antillean species. A, scanning electron micrograph of scale ("*Aleuracanthus* sp.," according to the label data) on which an undescribed species of *Acletoxenus* from Sumatra was found feeding (82×). B, Pineapple mealybug, *Dysmicoccus brevipes*, on *Ctenanthe oppenheimeriana* ? (photo by R. J. Gill, California Department of Foods and Agriculture). This is the reported prey of several *Pseudiastata* species. C, *Polyporus tricholoma* (Polyporaceae: Basidiomycetes) bloom on Barro Colorado Island, Panama. White fleshy polypores like this serve as rendezvous sites for probably all species of *Zygothrica* and as breeding sites for some primitive species. D, *Heliconia mariae* (Heliconiaceae: Zingiberales) inflorescence. The fleshy bracts of zingiberaceous flowers are one of the primary breeding sites of *Zygothrica*. E, Black land crab, *Gecarcinus ruricola*, at El Yunque, Puerto Rico. *Drosophila carcinophila* larvae feed on the exudate in nephric grooves of this crab (photo by J. K. Liebherr). F, Drawing of a famous specimen in the NMNH. This is the maxilliped of *Gecarcinus ruricola* taken from a specimen on Montserrat in 1894 by H. G. Hubbard. Twenty-two puparia of *Drosophila carcinophila* adhere to the inner surface, and four at the lower right have eclosed.

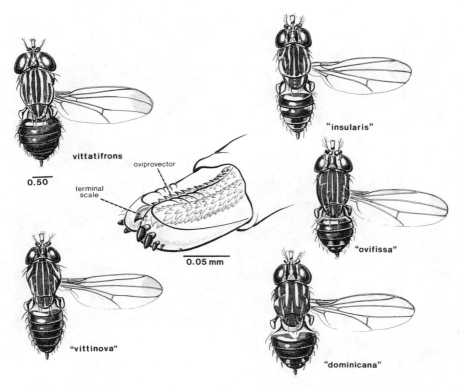

Figure 8-10. Insular Caribbean endemics of *Zygothrica* clades A+B (from Fig. 8-8B). An unusual feature that distinguishes this lineage is the possession of sharp, cleaverlike scales on the apex of an eversible ovipositor membrane (the oviprovector). Some species have apical wing spots that have become coalesced into the costal and subcostal wing cells (Z. "*dominicana*," Z. "*ovifissa*").

groups have contrasting life styles: cactophilics, crab commensals, homopteran predators, flower breeders, and mycophilous forms are represented on the Antilles (Fig. 8-9). Because of such ecological diversity, any biogeographic patterns are probably due more to historical factors than to a similarity in habits.

George Gaylord Simpson (1956:12) wrote: ". . . waifs by this route would be highly improbable, but they would be possible, and that is all that the theory [of dispersal] demands." If chance is the currency of dispersal sweepstakes, then it seems impossible that most of the Caribbean drosophilid lineages could have founded the younger mainland lineages via dispersal. At the very least, the islands are relatively depauperate in virtually every biological respect, and therefore any flow of colonists should be from the opposite (mainland to island) direction. A vicariant origin of the plesiomorphic Carib-

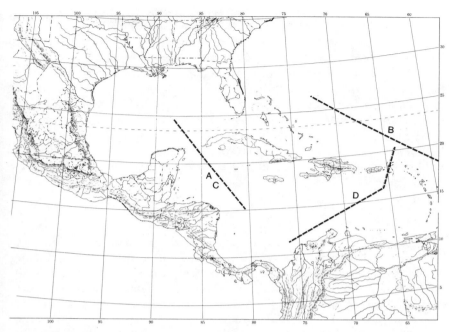

Figure 8-11. Hypothesis of vicariant events based on the distributions of Drosophilidae discussed in the text. A is the earliest event and represents a presumed separation of the Greater (and Lesser?) Antilles + Africa from Central America. B is the next event and is the separation of the West Indies from the west coast of Africa. C is an event involving the same land masses as in A, but at a later date and with a different type of geologic event (i.e., separation of a land bridge). D is the latest event and represents a separation of the Lesser and Greater Antilles, and of the Greater Antilles from northern South America.

bean drosophilid clades may require some equally complicated hypotheses, for example, one that assumes that ranges were at one time contiguous (i.e., that the islands have drifted or that they fused either geologically or during periods of low sea level). Using the approach established by Platnick and Nelson (1978), I present a hypothesis in Figure 8-11 for the Caribbean Drosophilidae which at least has geologic and biogeographic support.

Geologic evidence suggests that the Greater Antilles, and perhaps the northern Lesser Antilles, had a continental origin. Anderson and Schmidt (1983) have reviewed numerous studies and came to the conclusion that the Antilles formed from a shearing of several middle American plates beginning in the late Cretaceous (about 70 Ma). Durham (1985), whose evidence comes from marine fossils, has hypothesized that a large Caribbean plate moved through what was formerly the peninsula of southern Mexico and carried with it the portion of Central America that ranges from Honduras to Panama, Jamaica, southern Haiti, and the Lesser Antilles. This scenario apparently represents the geologic consensus on the origins of at least the Greater Antilles (reviewed in Rosen 1985). In my study I have been unable to resolve controversy over intraisland relationships (e.g., hybridization of Cuba and Hispaniola) because of the large areas of endemism. Durham (1985), however, has published the latest dates for early drifting, which are between 20 and 10 Ma. With North America, South America, and Africa joined into a predrift land mass, the Caribbean would fall at the point of their triple junction. Triple junctions are extremely active tectonically and are very likely to give contrasting dates of divergence, depending on what criteria are used. Finally, Donnelly (see chap. 2) supports the view that the Antilles not only evolved *in situ*, but were almost entirely submerged during periods in the Quaternary.

Fossils: Flies and Otherwise

Two drosophilid fossils were known prior to my recent study (Grimaldi 1987a): *Electrophortica succini* (Hennig 1965) (from Baltic amber, Eocene to early Oligocene [40–50 Ma]), and a *Neotanygastrella* (Chiapas amber, 30 Ma [Wheeler 1963]). The *Electrophortica* male possesses a pubescent arista and has a strong anterior reclinate seta, but it lacks a facial carina. Using these criteria, one can place the fossil basal to the *Pseudiastata* genus group, probably at an unresolved node from which *Pseudiastata* and *Acletoxenus* + *Mayagueza* originate. This also means that a primitive lineage (like *Pseudiastata*) formerly inhabited the Baltic region. *Neotanygastrella* is a distinctive group of flies with a distribution very similar to that of *Paramycodrosphila*, which I have already discussed. Of the 17 described species. 6 *Neotanygastrella* are Neotropical (mainland), 5 occur on the Ivory Coast, 4 are Indopacific insular, 1 is northern Australian, and another, *N. antillea*, occurs on Jamaica. *Neotanygastrella* is at least 30 million years old, but, given its Gondwanan distribution, it is probably even older. Dominican amber, which is at least early Miocene in origin (ca. 23 Ma), has thus far yielded 9 species of Drosophilidae (e.g., Fig. 8-12), representing 3 extant genera (*Chymomyza; Drosophila*, including 3 of its subgenera; *Scaptomyza*), and 2 extinct genera (Grimaldi 1987a). Because the genitalia could not be examined microscopically, it was not possible to place most of the species close to extant ones. It is clear, however, that the Drosophilidae were in existence when and where they could be affected by at least late Oligocene geophysical events in the Caribbean.

The abundance of fossils of animals other than flies makes extinction of Caribbean groups appear substantial. For ants, 22 of 37 genera and subgenera in Dominican amber remain on Hispaniola (Wilson 1985; see also chap. 9). Fifteen others have colonized the island since then, and 3 genera are extinct everywhere. Probably the most impressive account showing how

Figure 8-12. Female paratype of *Drosophila poinari* Grimaldi, fossilized in amber from the lower Miocene of the Dominican Republic (photo by G. O. Poinar, Jr.).

extinction can modify distribution is that of the Caribbean xerophilic vertebrates (Pregill and Olson 1981). Bats, owls, the extant mockingbird, *Mimus gundlachii* (on the Bahamas, Jamaica, and a fossil in Puerto Rico), caracaras, *Leiocephalus* and *Cyclura* iguanids, and meadowlarks (*Sturnella* spp.) are some taxa for which fossil evidence suggests that a wider distribution existed that usually encompassed the mainland. *Todus* (Todidae: Aves: Coraciiformes), which has 5 species, each endemic to a Greater Antillean island, once had a distribution that also included Wyoming in the Oligocene (Olson 1976). In fact, fossils show that the Todidae and another Caribbean group, the Momotidae (Coraciiformes), have a presently restricted distribution (Mourer-Chauvré 1982, Becker 1986). One figure speaks best: excluding bats, at least 75% of the Caribbean land mammals are extinct since at least the Pleistocene (Pregill and Olson 1981). At present, *Solenodon* is the largest native mammal on the Greater Antilles—a place where ground sloths (de Paula Couto 1967) and giant caviomorph rodents once roamed as well. Even though *Solenodon* is a terminal clade in the Solenodontoidea (MacFadden 1980), there is little doubt that the genus is relict because its endemicity is imposed by the extinction of close relatives—the Nesophontidae, Apternodontidae, and Geolabididae. Unfortunately, for most paleontologic studies there is too little evidence to allow a comparison between insular and mainland extinctions. For example, a single fossil locality on the mainland usually provides little data on the past distribution of the taxon to which the fossil belongs. Such a deposit would suggest a formerly wider distribution only when, first, there are taxa plesiomorphic to the fossil which lie within an area circumscribing the extant forms (e.g., within the Antilles); and second, the area has been examined for the presence of other fossils and extant taxa that show a similar biogeographic pattern, to ensure that the deposit is not an isolated case of ancient dispersal.

Other putatively relict lineages in the Caribbean are *Dugesia* flatworms (Ball 1971), *Polycentropus* caddisflies (Flint 1976), and some fishes (Rosen 1975). Among the ground beetles (Carabidae), *Antilliscaris* (Nichols 1986, 1987) and *Barylaus* (Liebherr 1986; see also chap. 6) are two genera known to have African and/or Malagasy affinities. In the Lepidoptera, groups showing Antillean-African connections are some Hesperiidae, the nymphalids *Calisto* and *Ypthima*, *Hypanartia* and *Antanartia*, and perhaps some species in the *Papilio thoas* species-group (Shields and Dvorak 1979). Wille and Chandler (1964) described *Trigona dominicana* (Apoidea: Meliponinae) from Dominican amber fossils. They considered it to be a primitive member of the subgenus *Hypotrigona*, most closely related to three species in tropical Africa and Madagascar, but Michener (1982) considers its affinities to be Central American. Five endemic Caribbean genera of Orthocladiinae Chironomidae (Diptera) have relatives that are either cosmopolitan (*Antillicladius, Lepurometriocnemus*), African-Holarctic (*Compterosmittia*), or Australian (*Petalocladius*) (Saether 1977, 1981, Sublette and Wirth 1972). Because few of the endemic chironomid species have a South American connection, Saether believes that vicariance and extinction have been biogeographically important in the Caribbean region. The genus *Wendilgarda* (Araneae: Theridiosomatidae) has three species: *W. mexicana* (Central America + Cuba), *W. clara* (Central America and northern South America), and *W. atricolor* (two small islands off the coast of Gabon, equatorial Africa) (Coddington 1986). Thus the African-Antillean connection is supported by ample data and fulfills predictions 2 and 4 by Rosen (1985:655). My review of these data does not necessarily imply that the Antillean fauna is an extremely ancient one, as may be the case for, say New Zealand or Tasmania. The fauna is certainly older than most people have previously realized, and the drosophilids are no exception in this regard. It is an important point to address since there is a common view that islands of continental origin harbor many relicts (Carlquist 1974), and because of the geologic controversy over the origins of the Antilles (see Donnelly, chap. 2).

It is speculative to interpret the past distributions of Caribbean Drosophilidae, but still compelling to consider why Antillean relicts may occur, at least for the fruit flies. Because of the extremely depauperate insular faunas, competitive release has been suggested dozens of time—and shown explicitly in some cases—as a reason for the success of some taxa. It is generally believed that insect populations are not limited by competition (Strong et al. 1984), but competition has been shown to be harsh in a guild of mycophagous *Drosophila* (Grimaldi and Jaenike 1984). In other words, relatively old drosophilid lineages on the islands may have an extended existence because, perhaps, of the availability of unoccupied niches. Jaenike (1978) concluded that competition and the genetic nature of peripheral populations probably account for the distribution of several closely related *Drosophila* species on islands off the coast of Maine.

A relaxed insular regime of competitive selection may account for the distribution in Florida of three generalist *Drosophila* species: *D. tripunctata*, and two species in the *D. cardini* group, *D. cardini* and *D. acutilabella*. *D. tripunctata* occurs in eastern North America and is the only nearctic member of the large (60 + spp.) *tripunctata* species-group (only two members of the group are Antillean endemics [Table 8-1], so this is an almost entirely Neotropical mainland group). Eight *cardini* group species are Antillean, and six are on the Central and South American mainland (Heed 1962), making this more of an insular group. At about the middle of the Florida peninsula, the replacement of the two *cardini* group species by *D. tripunctata* begins: in the Everglades, only an occasional *D. tripunctata* occurs among the many *cardini* group individuals; and at Gainesville (near the panhandle), only *D. tripunctata* can be found (J. Jaenike, pers. comm.). All three species come readily to the same bait, and their replacements appear independent of any obvious ecoclines (even temperature shows no such sharp change). Perhaps the *cardini* group flies, having evolved under insular conditions, cannot contend with a competitively superior mainland lineage. Population biologists interested in mechanisms of insular evolution would do well to do some comparative biology. In particular, the derived conditions of certain features considered important in colonization and demography, such as dispersal and competitive ability and host use, can be incorporated into a cladogram of a monophyletic group having mainland and insular species. Some species in *D. willistoni* or especially *cardini*, and *repleta* species groups would be good examples for study.

Summary

Fifty-eight species of Drosophilidae belonging to 9 genera are known only from the Antilles. Thirty-eight of them occur on the greater Antilles, with 20 species occurring on Jamaica, 16 on Puerto Rico, 12 on Hispaniola, and 9 on Cuba (the latter small number is due primarily to a lack of collecting). The relationships of three lineages are discussed in detail: *Mayagueza* (a monotypic genus found on Puerto Rico) and the two most speciose drosophilid taxa in the Caribbean and Neotropical regions, which are the *Drosophila repleta* species group and the genus *Zygothrica*. The closest relative of *Mayagueza* is *Acletoxenus*, a primarily southeast Asian genus that also has one Palearctic species. Insular endemics of the *repleta* group have their closest relatives widespread throughout Central America, and, in three of the four lineages, the insular lineages appear to have diverged before the formation of mainland taxa. This pattern is also seen in the *Zygothrica poeyi* species group, but the other main lineage of Antillean *Zygothrica* has an unusual countercurrent distribution along the Lesser Antillean arc. Hypothesized extinction of formerly widespread ancestors in some of the lineages may account for endemism, suggesting that the flies belong to relict lineages. The above three groups have contrasting life styles, which indicates that the biogeographic patterns are more an effect of historical rather than ecological factors. Ages of fossil Drosophilidae show that the family was developed enough to have been affected by at least by late Oligocene–early Miocene Caribbean tectonic movement. A hypothesis of four primary vicariance events for the Antillean Drosophilidae is presented.

Appendix 8.1: Endemic species of Antillean Drosophilidae
(classification following Wheeler 1981)

Distribution data come from Wheeler (1981), from specimens that were examined at the Smithsonian Institution, from the American Museum of Natural History, and from revisions.

Subfamily Steganinae
  Stegana
    (*Steganina*)
      horae Williston, 1896 (St. Vincent)
      tarsalis Williston, 1896 (St. Vincent)
  Mayagueza
      argentifera Wheeler, 1960 (Puerto Rico: Mayaguez)

Subfamily Drosophilinae
  Chymomyza
      microdiopsis Grimaldi, 1986 (Cuba: Santa Clara Prov.)
      jamaicensis Grimaldi, 1986 (Jamaica: Hardware Gap)
  Diathoneura
      metallica (Sturtevant), 1921 (Cuba: Bartle; Puerto Rico: Adjuntas; Dominican Republic: Cabo rojo)
  Drosophila
    (*Drosophila*)
    Calloptera species group
      ornatipennis Williston, 1896 (Cuba: throughout; Puerto Rico: Adjuntas, El Yunque, Maricao; Haiti: Jamaica, St. Vincent)

Canalinea species group
  paracanalinea Wheeler, 1957 (Puerto Rico: El Yunque, Rio Piedras)
Cardini species group
  acutilabella Stalker, 1953 (S. Florida; Cuba: Jamaica: Hispaniola)
  antillea Heed, 1962 (St. Lucia)
  arawkana Heed, 1962 (St. Kitts; Guadeloupe)
  bedicheki Heed & Russell, 1971 (Trinidad)
  belladunni Heed & Krishnamurthy, 1959 (Jamaica: Hardware Gap)
  caribiana Heed, 1962 (Martinique)
  dunni Townsend & Wheeler, 1955 (Puerto Rico: Rio Piedras; St. Thomas)
  nigrodunni Heed & Wheeler, 1957 (Barbados)
  similis Williston, 1896 (Cuba: Bartle; Grenada; Jamaica: St. Vincent)
Flavopilosa species group
  nesiota Wheeler & Takada, 1962 (Haiti: Petionville)
Tripunctata species group
  mediodiffusa Heed & Wheeler, 1957 (Haiti; Jamaica; Puerto Rico: El Yunque; Cuba)
  spinatermina Heed & Wheeler, 1957 (Trinidad: Pt. of Spain)
Drosophila
  (*Drosophila*)
  Repleta species group
  repleta subgroup:
  peninsularis Patterson & Wheeler, 1942 (Cuba; S. Florida; Jamaica; Puerto Rico)

  mulleri subgroup: *stalkeri* complex
  stalkeri Wheeler, 1954 (Cayman Is.; S. Florida; Jamaica)
  richardsoni Vilela, 1983 (Dominica; Montserrat; Puerto Rico: La Paraguera, Mayaguez; Virgin Gorda*)

  mulleri subgroup: *eremophila* complex
  n.sp. (Dominican Republic; Haiti; Jamaica)

  mulleri subgroup: *mulleri* complex
  mayaguana Vilela, 1983 (Bahamas; Conception; Dominican Republic; Gr. Inagua; Gr. Cayman; Haiti; Mayaguana Is.; Tortola)
  n.sp. 1 (Haiti; Nauassa Is.*)
  n.sp. 2 (Dominican Republic; Haiti; Jamaica: Pt. Henderson*)

  fasciola subgroup:
  mojuoides Wasserman, 1962 (Trinidad)
  paraguttata Thompson, 1957 (Jamaica: Bath)

  mercatorum subgroup:
  carcinophila Wheeler, 1960 (Gr. Cayman; Montserrat; Puerto Rico: Mona Is., Cueva los Lirios)

  unplaced subgroup:
  ramsdeni Sturtevant, 1916 (Cuba: Guantanamo)

Unplaced to species group
  coffeata Williston, 1896 (St. Vincent)
  endobranchia Carson & Wheeler, 1968 (Cuba?; Gr. Cayman)
  verticis Williston, 1896 (St. Vincent)

  (*Hirtodrosophila*)
    pleuralis Williston, 1896 (St. Vincent)
    prognatha Sturtevant, 1916 (Puerto Rico: Adjuntas; San Domingo)
Drosophila
  (*Sophophora*)
  Saltans species groups
    lusaltans Magalhaes, 1962 (Puerto Rico)
    milleri Magalhaes, 1962 (Puerto Rico: El Yunque)
    pulchella Sturtevant, 1916 (Montserrat; St. Vincent)
  Willistoni species group
    insularis Dobzhansky, 1957 (St. Kitts: St. Lucia)
    neoalagitans Wheeler & Magalhaes, 1962 (Haiti: Petitionville, Kenscoff; Jamaica: Hardware Gap)
  Unplaced to subgenus
    fusca Coquillett, 1900 (Puerto Rico: Utado)
    illota Williston, 1896 (St. Vincent)
    lutea (Wiedemann), 1830 (not specified)
    sororia Williston, 1896 (St. Vincent)
Neotanygastrella
  antillea Wheeler, 1957 (Jamaica: Montego Bay)
Paraliodrosophila
  antennata Wheler, 1957 (Jamaica: Bath)
Paramycodrosophila
  nephelea Wheeler, 1968 (Jamaica: Hermitage, Windsor)
Zygothrica
    **dominicana* (Dominica)
    **hypophallus* (Dominica)
    **insularis* (Dominica; Montserrat; St. Kitts; St. Vincent; Trinidad)
    **ovifissa* (Haiti: Kenscoff; Jamaica)
    **palpivanna* (Jamaica: Hermitage)
    **parasemistriata* (Haiti: Kenscoff)
    poeyi (Sturtevant), 1916 (Cuba: Havana; Dominican Republic: Las Abejas; Puerto Rico: Rio Piedras, Sabana)
    **vitrea* (Jamaica: Hardware Gap)
    vittatifrons (Williston), 1896 (St. Vincent; St. Kitts)
    **vittinova* (Jamaica: Hermitage; Puerto Rico: Cidra)

---

* Marvin Wasserman, pers. comm., June 1985
** are undescribed species (Grimaldi ms.)

Appendix 8.2: Taxonomic Characters Used in the Phylogenetic Hypothesis of Some Steganinae (Fig. 8-3)

"A" is the apomorphic and "P" the plesiomorphic state of each character.

Adults

1. Genitalia, male.   A: complex, hypandrium reduced to narrow arch; sternite X with complex ventroapical process.
P: hypandrium resembles shape of more proximal sternites; sternite X without processes; unadorned gonites.
2. Face.   A: carinate.   P: flat.
3. Arista (flagellomere 3).   A: pubescent.   P: short or long plumose.
4. Genitalia, male.   A: aedeagus lost, or reduced to very small endophallus. P: aedeagus robust.
5. Prescutellar setae.   A: 2 or more pairs present.   P: 1 pair.
6. Clypeus.   A: bulbous; height at least one-half of width; protrudes in profile. P: flat, barely visible in lateral view.
7. Leg coloration.   A: tibiae with 2 dark bands each.   P: tibiae unicolorous light.
8. Intraocellar setulae.   A: lost.   P: 1–3 (usually 2) pair present.
9. Third costal wing vein section.   A: possesses blunt spines.   P: blunt spines absent.
10. Spermathecal capsule.   A: possesses papillae.   P: glabrous.
11. Face.   A: short, length much less than length of front of head.   P: face about equal in length to frontal region.
12. Inner vertical setae.   A: parallel, directed backward.   P: convergent.
13. Ocellar setae.   A: absent/greatly reduced.   P: robust, length ≥ length of postocellars.
14. Spermathecal capsule.   A: glabrous, spherical, weakly sclerotized.   P: large, heavily sclerotized, with annulations or papillae.
15. Facial and frontal regions.   A: extremely narrow (width approximately equal to that of outer ocelli), inner eye margins parallel or nearly so.   P: width is ≥ 3× width of outer ocelli.
16. Ventral epandrial lobe (male).   A: elongate, with sclerotized and pointed end (toothlike).   P: simple, rounded lobe.
17. Ocellar setae.   A: convergent/cruciate.   P: divergent.
18. Setae coloration.   A: golden.   P: black.
18a. Interfrontal setulae.   A: numerous, cover almost entire front, 50 or more in number.   P: setulae on anterior portion of front, near ptilinum; 25 or less in number.

Larvae/Puparia (Instar III)

19. Cephalopharyngeal skeleton: epipharyngeal sclerite.
A: absent/greatly reduced (fused to tentorial phragma).
P: parastomal bar distinct, with latticed process.
20. Anterior spiracles.   A: reduced filament length (length ≤ width), budlike.
P: prominent spiracular filaments (length > 2× width).

21. Thoracic and abdominal spicules. A: absent (lost). P: present.
22. Cephalopharyngeal skeleton: sclerotization. A: only anterior portion of mandible sclerotized. P: most of skeleton sclerotized, including all of mandibles and anterior bridges plus cornuas.
23. Thoracic and abdominal spicule shapes. A: bifid. P: single point.
24. Structure of spiculeless surface. A: poroid, (waxy?) exudate. P: smooth.
25. Cephalopharyngeal skeleton. A: hypopharyngeal sclerite fused to tentorial phragma. P: sclerite detached, but articulates with, tentorial phragma.

Acknowledgments

I am indebted to three people in the preparation of this paper: Marvin Wasserman, who provided me with up-to-date information on the *repleta*-group flies; Wayne Mathis, who has been extremely supportive in my use of the Smithsonian collection; and James Liebherr, who inspired some of this work by inviting me to participate at the symposium held in Hollywood, Florida. Kevin Barber kindly provided me with larval specimens of *Rhinoleucophenga*, John Jaenike shared his unpublished data on Florida collections of *Drosophila*, and William Heed provided further data on Caribbean *Drosophila*. William Heed, Wayne Mathis, Toby Schuh, and Jim Liebherr kindly provided critical input to the manuscript, which Carol Ievolella skillfully typed and edited. Some of this work was done while I was a Griswold Fellow in the Entomology Department at Cornell and supported in part by a National Science Foundation doctoral dissertation grant.

References

Anderson, T.H., and V.A. Schmidt. 1983. The evolution of Middle America and the Gulf of Mexico–Caribbean Sea Region during Mesozoic time. Geol. Soc. America Bull. 94:941–966.

Ashburner, M. 1981. Entomophagous and other bizarre Drosophilidae. *In* M. Ashburner et al., eds., The Genetics and Biology of *Drosophila*, vol. 3a, pp. 395–429. Academic Press, New York.

Ball, I.R. 1971. Systematic and biogeographical relationships of some *Dugesia* species (Trichladida, Paludicola) from Central and South America. Amer. Mus. Novit. 2472:1–25.

Barker, J.S.F., and W.T. Starmer, eds. 1982. Ecological Genetics and Evolution. Academic Press, Sydney.

Becker, J.F. 1986. A fossil motmot (Aves: Motmotidae) from the Late Miocene of Florida. Condor 88:478–482.

Burla, H. 1956. Die Drosophiliden-Gattung *Zygothrica* und ihre Bieziehung zur *Drosophila*-Untergattung *Hirtodrosophila*. Mitt. Zoolog. Mus. Berlin 32:190–321.

Carlquist, S. 1974. Island Biology. Columbia University Press, New York. 660 pp.

Carson, H.L. 1967. The association between *Drosophila carcinophila* Wheeler and its host, the land crab *Gecarcinus ruricola* (L.). Amer. Midland Naturalist 78:324–343.

———. 1974. Three flies and three islands: Parallel evolution in *Drosophila*. Proc. Nat. Acad. Sciences (USA) 71:3517–3521.

Coddington, J.A. 1986. The genera of the spider family Theridiosomatidae. Smithsonian Contrib. to Zoology 422:1–96.

de Paula Couto, C. 1967. Pleistocene edentates of the West Indies. Amer. Mus. Novit. 2304:1–55.

Dobzhansky, T. 1957. Genetics of natural populations, XXVI: Chromosomal variability in island and continental populations of *Drosophila willistoni* from Central America and the West Indies. Evolution 11:280–293.

Durham, J.W. 1985. Movement of the Caribbean plate and its importance for biogeography in the Caribbean. Geology 13:123–125.

Eldredge, N., and S. Stanley, eds. 1984. Living Fossils. Springer-Verlag, New York. 291 pp.

Flint, O. 1976. The Greater Antillean species of *Polycentropus* (Trichoptera, Polycentropidae). Proceedings of the Biological Society of Washington 89:233–246.

Gonçalves, C.R. 1939. Biologia de Uma *Pseudiastata* depredadora de *Pseudococcus brevipes* (Dip. Diastatidae). Physis 17:103–112.

Grimaldi, D.A. 1986. The *Chymomyza aldrichii* species-group (Diptera: Drosophilidae): Relationships, new Neotropical species, and the evolution of some sexual traits. J. New York Entomol. Soc. 94:342–371.

———. 1987a. Amber fossil Drosophilidae (Diptera), with particular reference to the Hispaniolan taxa. Amer. Mus. Novit. 2880:1–23.

———. 1987b. Phylogenetics and taxonomy of *Zygothrica* (Diptera: Drosophilidae). Bull. Amer. Mus. Nat. Hist. 186:103–268.

Grimaldi, D., and J. Jaenike. 1984. Competition in natural populations of mycophagous *Drosophila*. Ecology 65:1113–1120.

Heed, W.B. 1962. Studies in the genetics of *Drosophila*, IX: Genetic characteristics of island populations. University of Texas Publications 6205:174–206.

Hennig, W. 1965. Die Acalyptratae des Baltischen Bernsteins und ihre Bedeutung fur die Erforschung der phylogenetischen Entwicklung dieser Dipteren-Gruppe. Stuttgarter Beitr. Naturkunde 145:1–214.

Jaenike, J. 1978. Ecological genetics in *Drosophila athabasca*: Its effect on local abundance. American Naturalist 112:287–299.

Liebherr, J.K. 1986. *Barylaus*, new genus (Coleoptera: Carabidae) endemic to the West Indies with Old World affinities. J. New York Entomol. Soc. 94:83–97.

McAlpine, J.F. 1968. An annotated key to drosophilid genera with bare or micropubescent aristae and a revision of *Paracacoxenus* (Diptera: Drosophilidae). Canadian Entomologist 100:514–532.

Màca, J. 1977. Revision of palaearctic species of *Amiota* subg. *Phortica* (Diptera, Drosophilidae). Acta Entomologica Bohemoslavaca 74:115–130.

———. 1980. Czechoslovak Drosophilidae (Diptera) with micropubescent arista. Acta Universitatis Carolinae—Biologica 1977:337–342.

———. 1980b. European species of the subgenus *Amiota* s.str. (Diptera, Drosophilidae). Acta Entomologica Bohemoslavaca 77:328–346.

MacFadden, B.J. 1980. Rafting mammals or drifting islands? Biogeography of the greater Antillean insectivores *Nesophontes* and *Solenodon*. J. of Biogeography 7:11–22.

Michener, C.D. 1982. A new interpretation of fossil social bees from the Dominican Republic. Sociobiol. 7:37–45.

Mourer-Chauvré, C. 1982. Les oiseaux fossiles des Phosphorites du Querey (Éocène supérieur à Oligocene supérieur): Implications paléobiogéographiques. Geobios (Lyon), Mémoire Spécial 6:413–426.

Nichols, S.W. 1986. Descriptions of larvae of Puerto Rican species of *Antilliscaris* Banninger and notes about relationships and classification of *Antilliscaris* (Coleoptera: Carabidae: Scaritini: Scaritina). Coleopterists Bull. 40:301–311.

Okada, T. 1968. Systematic Study of the Early Stages of Drosophilidae. Bunka Zugeisha Co., Ltd. Tokyo. 188 pp.

———. 1976. Subdivision of the genus *Chymomyza* Czeryny [sic] (Diptera, Drosophilidae), with description of three new species. Kontyû 44:496–511.

Olson, S.L. 1976. Oligocene fossils bearing on the origins of the Todidae and the Momotidae (Aves: Coraciiformes). *In* S.L. Olson, ed., Collected Papers in Avian Paleontology Honoring the 90th Birthday of Alexander Whetmore, pp. 11–119. Smithsonian Contrib. to Paleobiol. 27.

Platnick, N.I., and G. Nelson. 1978. A method of analysis for historical biogeography. Syst. Zool. 27:1–16.

Pregill, G.K., and S.L. Olson. 1981. Zoogeography of West Indian vertebrates in relation to Pleistocene climatic cycles. Ann. Rev. Ecol. and Systematics 12:75–98.

Rosen, D.E. 1975. A vicariance model of Caribbean biogeography. Syst. Zool. 24:431–464.

———. 1985. Geological hierarchies and biogeographic congruence in the Caribbean. Ann. Missouri Bot. Gardens 72:636–659.

Saether, O.A. 1977. Female genitalia in Chironomidae and other Nematocera: Morphology, phylogenies, keys. Bull. Fisheries Res. Board Canada 197:1–209.

———. 1981. Orthocladiinae (Diptera: Chironomidae) from the British West Indies, with descriptions of *Antillocladius* n.gen., *Lipurometriocnemus* n.gen., *Compterosmittia* n.gen. and *Diplosmittia* n.gen. Entomologica Scandanavica, supp. 16:1–46.

Shields, O., and S.K. Dvorak. 1979. Butterfly distribution and continental drift between the Americas, the Caribbean and Africa. J. Natural Hist. 13:221–250.

Simpson, G.G. 1956. Zoogeography of West Indian mammals. Amer. Mus. Novit. 1759:1–28.

Strong, D.R., J.H. Lawton, and Sir R. Southwood, eds. 1984. Insects on Plants: Community Patterns and Mechanisms. Harvard University Press, Cambridge, Mass.

Sublette, J.E., and W.W. Wirth. 1972. New genera and species of West Indian Chironomidae (Diptera). Florida Entomologist 55:1–17.

Throckmorton, L.H. 1975. The phylogeny, ecology, and geography of *Drosophila*. *In* R.C. King, ed., Handbook of Genetics, vol. 3, pp. 421–469. Plenum, New York.

Vilela, C.R. 1983. A revision of the *Drosophila repleta* species group (Diptera, Drosophilidae). Revista Brasiliera Entomologia 27:1–114.

Wasserman, M. 1982. Cytological evolution in the *Drosophila repleta* species group. *In* J.S.F. Barker and W.T. Starmer, eds., Ecological Genetics and Evolution, pp. 49–64. Academic Press, Sydney.

Wheeler, M.R. 1960. A new genus and two new species of neotropical flies (Diptera; Drosophilidae). Entomol. News 21:207–213.

———. 1963. A note on some fossil Drosophilidae (Diptera) from the amber of Chiapas, Mexico. J. Paleontology 37:123–124.

Wille, A., and L.C. Chandler. 1964. A new stingless bee from the Tertiary amber of the Dominican Republic (Hymenoptera: Melipononi). Rev. Biol. Trop. 12:187–195.

Wilson, E.O. 1985. Invasion and extinction in the West Indian ant fauna: Evidence from the Dominican amber. Science 229:265–267.

9 · The Biogeography of the West Indian Ants (Hymenoptera: Formicidae)

Edward O. Wilson

By my count, 89 genera and well-marked subgenera (see footnote in Table 9-1) comprising 383 species of ants have been recorded from the West Indies. Three of the genera (*Codioxenus, Dorisidris, Hypocryptocerus*), constituting 3.4% of the total, are endemic in the special sense of being known so far only from the West Indies; each is also limited to a single island. A total of 176 species, or 46% of the entire fauna, is known solely from the West Indies. Of these, 154, or 40.2% of the entire fauna, have been recorded from a single island—in other words, they might be insular endemics in the strict sense.

Our knowledge of this large and diverse fauna has grown to the point that several interesting generalizations can be made about its origin and dispersal. At the same time, substantial gaps remain, leaving open the possibility for significant discoveries in the future. In particular, most of the genera are in serious need of taxonomic revision, while some of the smaller and medium-sized islands remain largely unexplored. In addition, large collections of ants in the Tertiary (late Oligocene or early Miocene) amber of the Dominican Republic have recently become available, and they are yielding a new and in some cases surprising view of the history of the Greater Antillean fauna.

The purpose of this article is to provide a synopsis of the West Indian ant fauna in terms that can be utilized by biogeographers, while calling attention to some of the principal research opportunities that remain. The islands included in my survey of the living fauna are the Bahamas, the major arc of the Antilles from Cuba to Grenada, and Trinidad and Tobago. I have omitted Margarita, Blanquilla, the Roques archipelago, and the Dutch West Indies (Aruba, Curaçao, Bonaire), partly because they have westward locations well away from the Orinoco delta and hence are less likely to have served as steppingstones to the Lesser Antilles, and partly because their ant

Table 9-1. Diversity and endemicity in the ant faunas of the better-known West Indian islands

Island	Area (km²)	Genera and Subgenera[1]				Species			
		No. of nonendemic genera	No. of endemic genera	Total no. of genera	Percentage generic endemicity	No. of nonendemic species	No. of endemic species	Total no. of species	Percentage species endemicity
Cuba	114,525	45	2	47	4.44	74	72	146	49.66
Bahamas	13,864	28	0	28	0	52	7	59	12.07
Jamaica	10,991	29	0	29	0	53	6	59	10.34
Hispaniola	76,190	36	1	37	2.70	57	31	88	35.63
Mona	52	11	0	11	0	13	0	13	0
Puerto Rico	8,897	31	0	31	0	52	4	56	7.27
Culebra	26	17	0	17	0	26	0	26	0
St. Thomas	83	26	0	26	0	37	2	39	5.13
St. Croix	218	6	0	6	0	7	0	7	0
Antigua	280	7	0	7	0	11	0	11	0
Montserrat	102	6	0	6	0	7	0	7	0
Guadeloupe	1,780	14	0	14	0	15	0	15	0
Dominica	728	22	0	22	0	28	0	28	0
Martinique	1,116	17	0	17	0	20	0	20	0
St. Lucia	616	12	0	12	0	15	1	16	6.67
St. Vincent	389	35	0	35	0	57	6	63	9.68
Barbados	430	18	0	18	0	16	2	18	11.76
Grenada	344	23	0	23	0	38	1	39	2.63
Tobago	300	15	0	15	0	18	0	18	0
Trinidad	4,828	60	0	60	0	126	22	148	14.86

[1] The following relatively well-marked subgeneric distinctions were included in these counts, preliminary to a clearer resolution of their status: *Crematogaster* s.s. and *C.* (*Orthocrema*); *Leptothorax* (*Macromischa*) and *L.* (*Nesomyrmex*); *Pachycondyla* s.s. and *P.* (*Nylanderia*); *Solenopsis* s.s., *S.* (*Diplorhoptrum*), and *S.* (*Euophthalma*). All these taxa together count for 12 "well-marked" subgenera equivalent to full genera; the genera to which they belong (e.g., *Crematogaster*) were *not* then added to the counts for the West Indies as a whole.

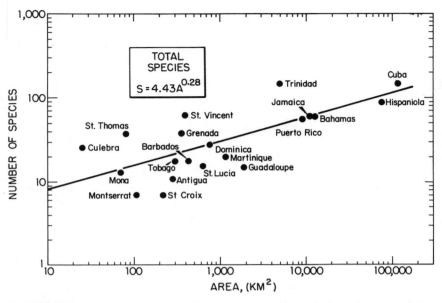

Figure 9-1. The relation between area and total species numbers in the better-known ant faunas of the West Indian islands. Curve estimated by least squares method.

faunas are so poorly known. Finally, I have included a brief account of our rapidly growing knowledge of the ants of Dominican amber.

The Data Base

Locality records of the living fauna were compiled earlier from the literature in catalog form by Kempf (1972). This extraordinarily useful publication has been supplemented by revisions of *Odontomachus* by Brown (1976), the ecitonine army ants by Watkins (1976), *Leptothorax* (*Macromischa*) by Baroni Urbani (1978), and portions of *Pheidole* by Brown (1981), as well as a synopsis of the Cuban ant fauna by Alayo (1974). I have recently provided updated lists of the genera of Haiti and the Dominican Republic (Wilson 1985a–d). Earlier species-level treatments of individual islands include Forel's study (1893) of the unusually thorough collections made by H. H. Smith on St. Vincent, as well as monographs and supplementary studies on Cuba by Wheeler (1913), Hispaniola by Wheeler and Mann (1914) and Wheeler (1936), and Puerto Rico by Smith (1936) and Torres (1984). Levins et al. (1973) made thorough collections of ants on 140 islands and cays of the Puerto Rico bank, plus the nearby islands of St. Croix, Mona, Monito, and Desecheo. They provided a number of interesting ecological

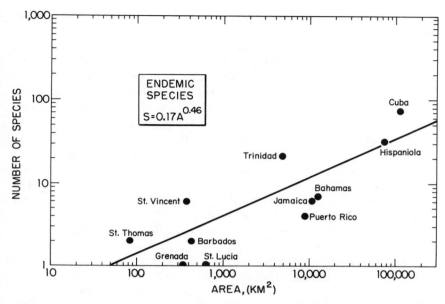

Figure 9-2. The relation between area and numbers of endemic species in the better known ant faunas of the West Indian islands. In some cases, especially the smaller islands, the numbers probably represent overestimates, due to the likelihood that species now known only from a single island (hence counted as "endemic") will eventually be found elsewhere. Curve estimated by least squares method.

and biogeographic generalizations concerning the 52 species encountered, but unfortunately did not publish an island-by-island list (the material is deposited in the Museum of Comparative Zoology, Harvard University).

Additional, unpublished collection data used in the present study, originating especially from Cuba, Hispaniola, and Trinidad, have been obtained by W. L. Brown and me. Current information on the ants of Dominican amber has been summarized in Baroni Urbani and Saunders (1982) and Wilson (1985a–e).

The Relation between Area and Diversity

The relations of area to numbers of genera and species, respectively, on the better-known islands are summarized in Table 9-1 and Figures 9-1 and 9-2. When one takes into account the fact that much of the variance is due to the still very imperfect collecting on most of the islands, the area-diversity curves are seen to be conventional. In particular, the faunistic exponent z (in $S = bA^z$, where S = number of species, A = island area, and b = a taxon area-specific constant) is 0.28, a typical number for closely grouped islands

(MacArthur and Wilson 1967). It is reasonable to suggest that the species numbers are at or near equilibrium, although this cannot be proved entirely from the area-diversity curves. Also, as depicted in Figure 9-2, the slope of the endemic species against area is steeper ($z = 0.46$, with few or no endemics being the rule in islands with areas under 1,000 km². Finally, Trinidad is well above the line of the main body of West Indian islands in both its absolute numbers of species and its absolute number of endemic species, but below the lines (see the far right column in Table 9-1) in *percentage* of endemic species. These disparities seem clearly due to Trinidad's origin as a continental island and its connection to the South American mainland during recent geologic times. Its fauna contains no fewer than 17 genera and well-marked subgenera that are widespread in South America but absent in the remainder of the West Indies. (These are *Acanthognathus*, *Apterostigma*, *Basiceros*, *Daceton*, *Dendromyrmex*, *Dolichoderus s.s.*, *Eciton*, *Ectatomma*, *Lachnomyrmex*, *Megalomyrmex*, *Oligomyrmex*, *Pachycondyla* (*Mesoponera*), *Pachycondyla* (*Neoponera*), *Procryptocerus*, *Talaridris*, *Tranopelta*, and *Zacryptocerus s.s.*, i.e. *Z. clypeatus*).

However, while these generalizations appear qualitatively robust, that is, unlikely to be altered in main form by future research, a strong caveat is needed. The West Indian ants are still poorly sampled in comparison with, say, birds, reptiles, and butterflies. Some islands, such as the Caymans, Barbuda, Anguilla, and St. Kitts, are virtually unexplored. Furthermore, while larger islands such as Cuba, Hispaniola, and Trinidad have been better worked, it is safe to predict that visits to remote areas, especially mountain forests, and a generous use of berlese collecting will yield many more species. The resulting general trend will probably be an overall rise in the area-species curve without much change in the slope. On the other hand, the ultimate effect on the curve relating area to the number of endemic species is difficult to predict. Many new endemic species will probably be discovered, especially the rarest and most locally distributed elements, but many other species now known from a single island (hence counted as "endemic") will turn up elsewhere. The processes of discovery are obviously antagonistic—but to a still unknown degree.

With the exception of Trinidad, which is little more than an extension of South America, the Antillean islands have faunas displaying the key traits indicative of a "sweepstakes" origin. In addition to the 17 genera that just reach Trinidad, other dominant elements of the New World fauna extend only to one or a few of the Lesser Antilles, in the great majority of cases to the lower arc of these islands. Examples include *Azteca*, the most abundant and diverse of the Neotropical dolichoderine genera, and other genera that are common and widespread in South and Central America, including *Cephalotes*, *Hypoclinea*, *Leptothorax* (*Nesomyrmex*), *Monacis*, *Myrmicocrypta*, and *Prionopelta*. Moreover, genera and species distributed continuously or near-continuously within the Lesser Antilles show various degrees of penetration northward past Trinidad into the lower arc of islands. Thus *Leptothorax* (*Nesomyrmex*) just reaches Grenada, *Neivamyrmex* reaches Grenada and just beyond, to St. Vincent, and so forth.

A second trait typical of a sweepstakes origin is a disharmonic composition: endemic species clusters occur on the larger islands, which in the case of Cuba and Hispaniola make up a disproportionate share of the native faunas. There are at least four such clusters: the *sphaericus* group or subgenus *Manniella* of *Camponotus* in Cuba (*micrositus, sphaericus, torrei*); the *gilviventris* group or subgenus *Myrmeurynota* of *Camponotus*, which also occurs on the mainland from South America to Mexico, with two species in Cuba (*gilviventris, thysanopus*) and five in Hispaniola (*albistramineus, altivagans, augustei, christophei, toussainti*)[1]; the *Rogeria brunnea* group of Cuba (*brunnea, caraiba, cubensis, scabra*); and *Leptothorax* (*Macromischa*) in all of the Greater Antilles except Jamaica. One group requires qualification: Charles Kugler (pers. comm.), who is currently revising *Rogeria*, considers the Cuban forms still to be of uncertain status even though they are quite likely a cluster of sibling species.

Leptothorax (*Macromischa*) has attained the status of a truly radiated group, as shown in the recent analysis by Baroni Urbani (1978). Cuba has no fewer than 38 species, constituting 26% of the entire known ant fauna. All but one are endemic to the island, and the single exception, *L. androsanus*, is shared only with the nearby Bahamas. As exemplified in Figure 9-3, the morphological diversity of the Cuban species is great enough to encompass what might well be recognized as several distinct genera in other parts of the world. Baroni Urbani believes that this fauna is in fact polyphyletic, with the picture being complicated further by convergence in the key traits of elongation and thickening of the femora and either the enlargement or the loss of the propodeal spines. The elongation of the appendages is correlated with an increase in overall size and thickening of the femora, all of which appear to improve the ant's ability to sting severely from an unusual posture: the gaster is bent all the way beneath the body and the large sting is projected forward. The elongated petiole clearly facilitates this movement, while stronger tibial levator muscles, accommodated by the swollen femora, lend the ants greater stability when they raise their bodies on their legs. Some of the species are metallic blue or green in color (or even gold in certain reflections of light) and can be seen foraging in conspicuous files during daylight hours. Considering the stinging ability of the workers, it is possible that the color is aposematic.

With reference to *Leptothorax* (*Macromischa*) as a whole, Baroni Urbani has concluded that at least two main evolutionary centers exist. One is Cuba,

[1]The "subgenera" of *Camponotus* noted here are of such ambiguous status that they were not included in the genera-plus-subgenera counts of Table 9-1 and elsewhere.

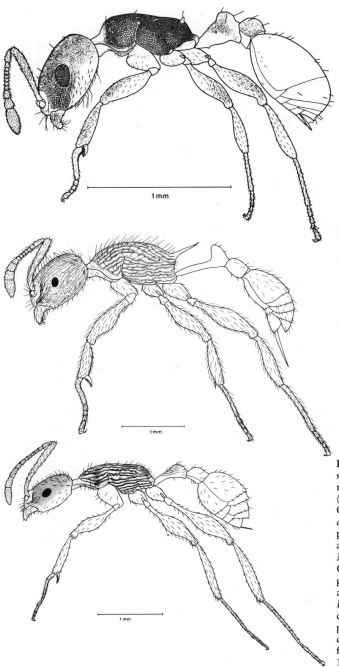

Figure 9-3. Representative species from the adaptive radiation of *Leptothorax* (*Macromischa*) from the Greater Antilles. *Upper*: *L. creolus* of the Dominican republic, a relatively generalized leptothoracine ant. *Middle*: *L. platycnemis*, a Cuban species with elongated legs, propodeal spines and petiolar peduncle. *Lower*: *L. iris*, a Cuban species with elongated legs, lost propodeal spines, and reduced petiolar node. (Modified from Baroni Urbani 1978.)

the species of which are adapted to live in the soil or in crevices of limestone. Another is southern Mexico and Guatemala, which contains arboreal species that often nest in orchids and other epiphytes. The Cuban radiation has spilled over into Hispaniola, with five endemic species, and also into Puerto Rico, which possesses three endemic species and another (*albispina*) shared with nearby Mona. It is surely one of the chief mysteries of West Indian biogeography that not a single species, endemic or otherwise, has been discovered on Jamaica.

The three endemic genera of the West Indies are each limited to a single species. *Codioxenus simulans* and *Dorisidris nitens* are small, inconspicuous ants known only from several collections in Cuba. They belong to the tribe Dacetini, a group of hypogaeic ants specialized (so far as known) as predators on collembolans and other soft-bodied arthropods. *Hypocryptocerus haemorrhoidalis*, a relatively weakly demarcated member of the exclusively arboreal tribe Cephalotini, is common and widespread on Hispaniola. (*Xenometra monilicornis*, previously thought to be an endemic of St. Thomas, is a synonym of *Cardiocondyla emeryi*, an Old World species introduced by human commerce into the West Indies; see Baroni Urbani 1973, and Kugler 1983.)

The Biogeographic Affinities of the Modern Fauna

The Antillean ant fauna can be divided into two sections with reference to geographic origin. The first comprises the species that have spread for varying distances from South America northward through the Lesser Antilles and, in some cases, on into the large islands of the Greater Antilles. Some of these populations are little changed across their ranges, while others have speciated to produce endemic forms on one or more of the islands. Examples include various species complexes within the genera *Acromyrmex, Anochetus, Atta, Azteca, Camponotus* (subgenera *Myrmaphaenus, Myrmobrachys, Myrmosphincta, Myrmothrix, Pseudocolobopsis,* and *Tanaemyrmex*), *Crematogaster* (subgenus *Orthocrema*), *Cyphomyrmex, Hypoclinea, Hypoponera, Iridomyrmex, Leptogenys, Monomorium, Mycetophylax, Mycocepurus, Myrmicocrypta, Neivamyrmex, Nesomyrmex* (usually placed as a subgenus of *Leptothorax*), *Octostruma, Odontomachus, Pachycondyla, Paracryptocerus, Pheidole, Platythyrea, Prionopelta, Pseudomyrmex, Smithistruma, Solenopsis* (subgenus *Diplorhoptrum*), *Strumigenys, Tapinoma, Trachymyrmex,* and *Wasmannia*.

A smaller group of species and complexes of species are each endemic to one or more of the middle-sized and large islands, especially the four major islands of the Greater Antilles (Cuba, Hispaniola, Jamaica, Puerto Rico). Examples include *Camponotus* (subgenera *Manniella* and *Myrmeurynota*),

Codiomyrmex, Codioxenus, Crematogaster (subgenus *Crematogaster*), *Ephebomyrmex, Gnamptogenys* (*schmitti* group), *Hypocryptocerus, Leptothorax* (subgenus *Macromischa*), *Myrmelachista, Paratrechina* (subgenus *Nylanderia*), *Rogeria* (*brunnea* group), and *Thaumatomyrmex*. Our imperfect knowledge of ant systematics, and of these groups in particular, preclude the identification of the ultimate geographic origin of the endemic Antillean species in the second group.

In addition to the two biogeographic categories just cited, there is at least one species that might have a North American origin: *Prenolepis gibberosa*, an endemic of the mountains of Cuba. Other known *Prenolepis* in the New World are *P. acuminata* from Jalapa on the eastern edge of the Mexican plateau and *P. imparis*, which is one of the most widespread of all ant species in the temperate regions of the United States, Canada, and Mexico. *Prenolepis* is an ancient genus, occurring in the early Oligocene Baltic amber. I have also recently found a specimen in the late Tertiary amber of the Dominican Republic.

The Dominican Amber Fauna

On the basis of the ant fauna, the West Indies today appear to be typical oceanic islands. By this I mean that a limited number of stocks colonized the island in the sweepstakes manner. Also, at least one group, *Leptothorax* (*Macromischa*), has undergone radiation so as to fill some of the vacant adaptive zones. But a strikingly different picture has emerged from the study of late Tertiary amber of the Dominican Republic, the inclusions of which are more characteristic of the contemporary mainland fauna of South and Central America, and particularly of the lowland humid tropical forests.

Earlier research by Baroni Urbani (1980a–d and in Baroni Urbani and Saunders 1982) on material in the Staatliches Museum für Naturkunde, Stuttgart, had disclosed the presence of the genera *Anochetus, Gnamptogenys, Paracryptocerus, Pseudomyrmex*, and *Trachymyrmex*. A poorly preserved set of workers in a single amber piece placed by Baroni Urbani (1980c) in *Leptomyrmex* has now been supplemented by better-preserved workers and a single male (Baroni Urbani and Wilson 1987).

In my own studies, based on 602 pieces of amber containing an estimated 1254 ants and located mostly in the Museum of Comparative Zoology, I have been able to expand this list greatly. The majority of the specimens came from the western amber-bearing region of the Dominican Republic, located in the mountainous La Cumbre region 10 to 20 km northeast of Santiago. A few came from the eastern amber-bearing region south of Sabana de la Mar. Still others, making up fewer than 5% of the specimens, are evidently younger than those from other Dominican localities. Schlee

(1984) placed their age as recently as 200 years, hence ranking them as copal rather than amber, but his evidence was circumstantial: the unusually clear, almost glasslike quality of the amber pieces and their location near the ground surface. In contrast, J. B. Lambert (pers. comm.) found that while the NMR spectra of Cotui material indicate a younger age than that of other Dominican amber, the deposits still seem to be Tertiary rather than Recent. Perhaps the Cotui material spans a substantial period of time, extending even into the Quaternary period. If so, more collecting and analysis might provide us with a unique opportunity to trace the island fauna in a more nearly continuous manner through time.

For purposes of analysis, the fossil and living genera can be conveniently classified into six biogeographic categories based on their presence or absence in the amber and the extent of their retreat since amber times (Table 9-2). Perhaps the single most interesting genus is *Neivamyrmex*, represented by *N. ectopus*, the first fossil army ant ever discovered (Wilson 1985b; see Fig. 9-4 here). No contemporary army ant is known from the northern arc of the Lesser Antilles or any of the Greater Antilles (Watkins 1976). Furthermore, army ants generally have very low colonizing ability across water. In the case of the New World Ecitoninae, *Neivamyrmex nigrescens* has been recorded from the Islas Marias, 100 km off the Mexican Pacific coast, while *N. klugi* occurs on St. Vincent in the lower arc of the Lesser Antilles. No

---

Table 9-2. The status of ant genera and well-marked subgenera on Hispaniola (Dominican Republic plus Haiti) (From Wilson 1985d)

Present in Dominican Amber

Now extinct worldwide: *Ilemomyrmex, Oxyidris,* new genus near *Rogeria*
Now extinct in Western Hemisphere: *Leptomyrmex*
Now extinct in the Greater Antilles but present elsewhere in the Neotropical region: *Azteca, Dolichoderus, Erebomyrma, Hypoclinea, Leptothorax (Nesomyrmex), Monacis, Neivamyrmex, Paraponera, Prionopelta*
Now extinct on Hispaniola but present elsewhere in the Greater Antilles: *Cylindromyrmex, Octostruma, Prenolepis*
Still present on Hispaniola: *Anochetus, Aphaenogaster, Camponotus, Crematogaster (Acrocoelia), Crematogaster (Orthocrema), Cyphomyrmex, Gnamptogenys, Hypoponera, Iridomyrmex, Leptothorax (Macromischa), Odontomachus, Pachycondyla (Trachymesopus), Paratrechina (Nylanderia), Pheidole, Platythyrea, Pseudomyrmex, Smithistruma, Solenopsis (Diplorhoptrum), Solenopsis (Solenopsis), Tapinoma, Trachymyrmex, Zacryptocerus*

New Arrivals

New World genera and well-marked subgenera present in the modern Hispaniolan fauna but unknown in the Dominican amber: *Acropyga, Brachymyrmex, Ephebomyrmex, Eurhopalothrix, Hypocryptocerus, Leptogenys, Monomorium, Mycocepurus, Myrmelachista, Solenopsis (Euophthalma), Strumigenys, Wasmannia* (possibly introduced by commerce)
Old World genera (or species groups within these genera) introduced within historical times by human commerce: *Cardiocondyla, Paratrechina (Paratrechina), Tetramorium*

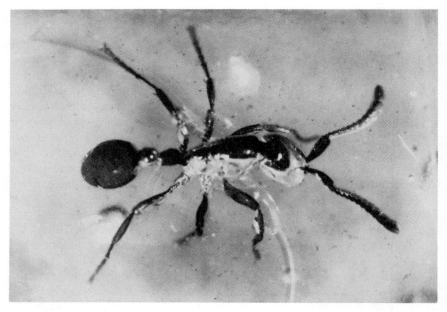

Figure 9-4. The extinct army ant *Neivamyrmex ectopus* from the amber of the Dominican Republic. (From Wilson 1985b.)

ecitonine is known farther away from the mainland. Similarly, the Old World Dorylinae, represented by *Aenictus*, extends only as far east as the Philippine Islands, New Guinea, and Queensland (Wilson 1964). It is wholly unknown from those portions of Micronesia and Polynesia that support a native ant fauna (Wilson and Taylor 1967). The farthest outlier in the western part of the range is a population of *A. fergusoni* on Great Nicobar Island, 160 km from Sumatra. The existence of *Neivamyrmex ectopus* in the Dominican amber is therefore consistent with the common view, based on both geologic and paleobotanical studies (Graham and Jarzen 1969), that the ancestral Antilles were larger and extended closer to the Mexican mainland during the Middle and Late Tertiary than is now the case. Furthermore, the overall closer similarity of *N. ectopus* to contemporary Mexican and United States species, as opposed to South and Central American species (Wilson 1985b), is consistent with a closer approach of the Greater Antilles to Mexico in particular than to the northern coast of South America during the Tertiary period. But this may be the only conclusion that can be drawn. The evidence, in my opinion, does not disclose whether the closer proximity was due to continental drift or merely to the temporary emergence of a larger Antillean land mass that extended westward.

Further evidence of a closer approach of Hispaniola to the mainland, however it occurred, is provided by the presence in the amber of *Dolichoderus, Monacis,* and *Paraponera,* which are today principal elements of moist tropical forests in South and Central America but do not extend north of Trinidad. (*Monacis bispinosus* has been cited from St. Thomas, but I consider this to be either an erroneous record or else based on an introduced population; see Wilson 1985c.) These genera, all comprising large, conspicuous ants, seldom reach offshore islands along the mainland coasts and hence appear to be limited in dispersal ability.

Another important fact revealed by the Dominican fossils is the remarkable retreat of the subfamily Dolichoderinae from the West Indies since Tertiary times. Four genera (*Azteca, Dolichoderus, Hypoclinea, Monacis*) have disappeared entirely from the Greater Antilles and one (*Leptomyrmex*) from the entire Western Hemisphere. Only two (*Iridomyrmex, Tapinoma*) have persisted to present times, while yet another genus, *Conomyrma,* has invaded more recently. The dominant species of the amber fauna was the now-extinct *Azteca alpha,* which, if it was like all modern *Azteca,* was arboricolous. According to W. L. Brown (pers. comm.), who has collected intensively in the Dominican Republic, the dominant arboricolous ants today include members of *Pseudomyrmex, Crematogaster, Zacryptocerus,* and *Camponotus.* In this important respect the West Indian fauna mirrors the general decline of the Dolichoderinae in North and South America, Europe, and Asia, possibly in conjunction with the advance of *Crematogaster* as a competitor of *Iridomyrmex* (Brown 1973).

The generic lists of the amber and modern Hispaniolan faunas can be summarized as follows. Of 38 genera and well-defined subgenera identified in the fossil deposits, 34 have survived somewhere in the New World tropics to the present, although the species studied thus far are extinct. Of the surviving genera and subgenera, 22 persist on Hispaniola. Fifteen genera and subgenera have colonized the island since amber times, restoring the number of genera and well-defined subgenera now present on Hispaniola to 37 (Wilson 1985a–e). It needs to be stressed that while the overall generic diversity has remained the same, the close numerical correspondence (38 and 37 genera, respectively) is surely only a coincidence, likely to be changed with further collecting of living and amber material.

Has the species diversity also stayed constant? The modern Hispaniolan fauna averages $88/37 = 2.35$ species per genus. The Hispaniolan amber fauna belonging to the Dolichoderinae and Ecitoninae, the only subfamilies revised so far to the species level, averages $11/8$ species $= 1.38$ per genus. If this ratio were to hold over all 38 genera, the amber fauna would contain 52 species rather than the 88 known from the living fauna. However, sample size represented by the two subfamilies is still too small to be certain that overall ant species diversity was less in amber times than today.

I have also conducted a search for biological traits of living species that were likely to have been shared with the Hispaniolan fossil members of the same genus and might have contributed to the extinction or survival of the genus on the island from Tertiary times (Wilson 1985d). The data were not numerous enough to support effects due to the flightless condition of the queen and hence reduced dispersal power of colonies caused by this trait, despite the clear-cut example of the *Neivamyrmex* army ants. Weak evidence, with confidence just below the 95% level (G test of independence, 2×2 tables), exists for an increase in extinction rate associated with large individual or colony size, biological traits that are likely to cause a reduction in the size of populations of colonies. Examples of genera in this condition include *Dolichoderus* and *Paraponera*. A positive relation also exists, in this instance above the 95% confidence level, between extinction and extreme specialization in either prey choice or nest site. Examples of such specialized genera and subgenera that became extinct on Hispaniola (or at least have not yet been found in the modern fauna there) include *Cylindromyrmex, Neivamyrmex, Octostruma*, and *Prionopelta*. Finally, genera and subgenera that occur in both the New World and Old World, and hence show evidence of greater colonizing ability, have also reached the Greater Antilles in greater numbers and become extinct less frequently on Hispaniola since amber times ($P < 0.05$). Examples include *Anochetus, Aphaenogaster, Camponotus (Tanaemyrmex), Crematogaster s.s., Crematogaster (Orthocrema), Gnamptogenys, Hypoponera, Iridomyrmex, Odontomachus, Paratrechina (Nylanderia), Pheidole, Platythyrea, Solenopsis (Diplorhoptrum), Tapinoma*, and *Trachymesopus*.

Discussion

Several enticing research opportunities have been more clearly revealed by the biogeographic analysis summarized here. First, some of the Antillean islands have never been carefully collected and are either sufficiently large in size, topographically varied enough, or strategically placed to be of more than passing interest. They include the Caymans, Isle of Pines, Barbuda, St. Kitts, Guadeloupe, Martinique, Dominica, Tobago, as well as Margarita and the other islands along the north coast of South America to the west of Trinidad. Jamaica still seems to me to be relatively undercollected among the Greater Antilles. And the more remote forested areas of Cuba and Hispaniola would certainly repay additional berlese sampling of the soil and litter.

Although Baroni Urbani has provided a valuable systematic monograph of *Leptothorax (Macromischa)*, very little has been learned concerning the biology of this most luxuriantly radiated of the West Indian ant groups. To what extent have the endemic forms, especially the 37 or more on Cuba, diversified in habitat selection, nest site, food habits, defensive behavior, colony size, and social organization? Has *Leptothorax* (*Macromischa*) undergone a typical radiation, in the sense of preferentially occupying niches left vacant by other ant genera that are still absent from the Greater Antilles?

Possibly linked to the *Macromischa* case is the mysterious absence of *Neivamyrmex* and other ecitonine army ants from the Greater Antilles. These insects are for the most part specialized predators on *Camponotus*, *Pheidole*, and other ants. In other islands lacking army ants, for example Fiji, New Caledonia, southern Australia, and Madagascar, cerapachyines such as *Cerapachys* and *Sphinctomyrmex* are disproportionately abundant. But the Cerapachyinae are also absent from the Greater Antilles, or nearly so; the only record to date is *C. biroi* (= *C. seini*), a southeast Asian species introduced into Puerto Rico, and this appears to be relatively scarce. So the question arises: What, if anything, specializes in ant predation in the Greater Antilles? Is it possible that some of the *Leptothorax* (*Macromischa*) species have filled this niche?

I find it also puzzling that the myrmicine genus *Pheidole* has not flourished to a greater extent in the Greater Antilles, despite the fact that it was present as far back as Dominican amber times. Elsewhere it is the dominant ant genus in warm temperate and tropical portions of the New World, with over 500 unchallenged names (and probably at least that many true species) and the greatest numbers of colonies and individual ants in most habitats. However, only 17 described species have been recorded from the Greater Antilles. Of these, a mere 8 are limited to these islands, and none show signs of even an early stage of adaptive radiation. On the other hand, a close study of *Pheidole* might well result in a challenge of my own subjective impression. W. L. Brown (pers. comm.), for example, considers the native *Pheidole* of the Dominican Republic, including several undescribed species he has collected, to be relatively abundant and diverse.

Summary

The known West Indian ant fauna comprises 383 species in 89 genera (including 12 well-marked subgenera; see footnote in Table 9-1). Three of the genera are known only from the West Indies, and in fact only from a single island. In addition, 176 species, or 46% of the total, have so far been recorded only from the West Indies. The highest endemicity is that of Cuba, with 49.7% of its 146 species having that status (see Table 9-1). The great majority of endemic Antillean species are found on islands with areas greater than 1000 km² (Fig. 9-2).

The area-species curve appears regular in form, with a coefficient of 0.28, a property that supports but does not prove species equilibria (Fig. 9-1). If Trinidad is excluded because of its status as a continental island close to the South American mainland, the Antillean fauna is disharmonic in the manner typical for oceanic islands. In particular, a number of dominant mainland ant genera, including *Azteca, Cephalotes*, and *Monacis*, are wholly lacking in the Greater Antilles. And one myrmicine group, *Leptothorax* (*Macromischa*), has undergone a dramatic adaptive radiation, producing at least 37 endemic species on Cuba alone.

The modern West Indian ant fauna can be divided into two principal groups with respect to origin. One set comprises species that have spread to varying degrees of penetration from South America northward through the Lesser Antilles and in some cases onto the larger islands of the Greater Antilles. The second set, which is smaller and appears generally older, comprises species that are endemic to one or more of the middle-sized to larger islands, especially the four major islands of the Greater Antilles (Cuba, Hispaniola, Jamaica, Puerto Rico). The ultimate origin of stocks in the second group cannot be determined on the basis of existing knowledge.

In contrast to the contemporary ants, the late Tertiary fauna of the Greater Antilles, as revealed by recent studies of the Dominican amber, is more nearly characteristic of a continental fauna. Several genera were present on Hispaniola, including *Leptomyrmex, Azteca, Dolichoderus, Paraponera*, and the army-ant genus *Neivamyrmex*, that are today abundant elsewhere but wholly absent in the modern Greater Antilles fauna. Because army ants especially are poor colonizers across water, it seems reasonable to conclude that the Greater Antilles were nearer to the mainland (and in particular to Mexico) during late Tertiary times than is the case today.

The Hispaniolan fauna has undergone considerable turnover since Dominican amber times while retaining constancy in the number of genera. Of 38 genera and well-defined subgenera identified in the amber, 34 have survived somewhere in the New World tropics to the present, although the species studied thus far are extinct. (Three genera are apparently entirely extinct: *Ilemomyrmex, Oxyidris*, and an undescribed genus near *Rogeria*.) Of the surviving genera and subgenera, 22 persist today on Hispaniola. Fifteen genera and subgenera have colonized the island since amber times, restoring the number now present on Hispaniola to 37.

A higher extinction rate occurred on Hispaniola in genera and subgenera that are either highly specialized or possess less colonizing ability as evidenced by their restriction to the New World. There is an indication that the same was true for genera with either large individual or colony size, which are biological traits often associated with smaller populations of colonies.

Acknowledgments

I am grateful to William L. Brown and Charles Kugler for advice and additional data provided during the course of this study. The research was supported by National Science Foundation Grant No. BSR–84–21062.

References

Alayo, P. 1974. Introduccion al estudio de los Himenopteros de Cuba. Superfamilia Formicoidea. Acad. Cienc. Cuba, Inst. Zool. 53:1–58.

Baroni Urbani, C. 1973. Die Gattung *Xenometra*, ein objektives Synonym (Hymenoptera, Formicidae). Mitt. Schweiz. Ent. Ges. 46(3–4):199–201.

———. 1978. Materiali per una revisione dei *Leptothorax* neotropicali appartementi al sottogenere *Macromischa* Roger, n. comb. (Hymenoptera: Formicidae). Ent. Basil. 3:395–618.

———. 1980a. First description of fossil gardening ants (Amber Collection Stuttgart and Natural History Museum Basel: Hymenoptera, Formicidae. I: Attini). Stutt. Beitr. Naturk. B 54:1–13.

———. 1980b. *Anochetus corayi* n.sp., the first fossil Odontomachiti ant (Amber Collection Stuttgart: Hymenoptera, Formicidae. II: Odontomachiti). Stutt. Beitr. Naturk. B 55:1–6.

———. 1980c. The first fossil species of the Australian ant genus *Leptomyrmex* in amber from the Dominican Republic (Amber Collection Stuttgart: Hymenoptera, Formicidae. III: Leptomyrmicini). Stutt. Beitr. Naturk. B 62:1–10.

———. 1980d. The ant genus *Gnamptogenys* in Dominican amber (Amber Collection Stuttgart: Hymenoptera, Formicidae. IV: Ectatommini). Stutt. Beitr. Naturk. B 67:1–10.

Baroni Urbani, C., and J.B. Saunders. 1982. The fauna of the Dominican amber: The present status of knowledge. Transactions of the 9th Caribbean Geological Conference (Santo Domingo, Dominican Republic) 1:213–223.

Baroni Urbani, C., and E.O. Wilson. 1987. The fossil members of the ant tribe Leptomyrmecini (Hymenoptera: Formicidae). Psyche 94(1–2):1–8.

Brown, W.L. 1973. A comparison of the Hylaean and Congo–West African rainforest ant faunas. *In* B.J. Meggers, E.S. Ayensu, and W.D. Duckworth, eds., Tropical Forest Systems: A Comparative Review, pp. 161–185. Smithsonian Institution Press, Washington, D.C.

———. 1976. Contributions toward a reclassification of the Formicidae. Part VI. Ponerinae, Tribe Ponerini, Subtribe Odontomachiti. Section A. Introduction, subtribal characters, genus *Odontomachus*. Studia Ent. 19(1–4):67–171.

———. 1981. Preliminary contributions toward a revision of the ant genus *Pheidole* (Hymenoptera: Formicidae), part I. J. Kansas Ent. Soc. 54(3):523–530.

Forel, A. 1893. Formicides de l'Antille St. Vincent, récoltés par Mons. H.H. Smith. Trans. Ent. Soc. Lond. 4:333–418.

Graham, A., and D.M. Jarzen. 1969. Studies in Neotropical paleobotany. I. The Oligocene communities of Puerto Rico. Ann. Missouri Bot. Gardens 56:308–357.

Kempf, W.W. 1972. Catálogo abreviado das formigas da Região Neotropical. Studia Ent. 5(1–4):3–344.

Kugler, J. 1983. The males of *Cardiocondyla* Emery (Hymenoptera: Formicidae) with the description of the winged male of *Cardiocondyla wroughtoni* (Forel). Israel J. Ent. 17:1–21.

Levins, R., M.L. Pressick, and H. Heatwole. 1973. Coexistence patterns in insular ants. Amer. Sci. 61:463–472.

MacArthur, R.H., and E.O. Wilson. 1967. The theory of island biogeography. Monog. Pop. Biol. No. 1. Princeton University Press, Princeton, N.J. 203 pp.

Schlee, D. 1984. Notizen über einige Bernsteine und Kopale aus aller Welt. Stutt. Beitr. Naturk. C 18:29–37.

Smith, M.R. 1936. The ants of Puerto Rico. J. Agric. Univ. Puerto Rico 20(4):819–875.

Torres, J.A. 1984. Niches and coexistence of ant communities in Puerto Rico: Repeated patterns. Biotropica 16(4):284–295.

Watkins, J.F., II. 1976. The Identification and Distribution of New World Army Ants. Baylor University Press, Waco, Texas. x, 102 pp.

Wheeler, W.M. 1913. The ants of Cuba. Bull. Mus. Comp. Zool. Harv. 54(17):477–505.

———. 1936. Ants from Hispaniola and Mona. Bull. Mus. Comp. Zool. Harv. 80(2):195–211.

Wheeler, W.M., and W.M. Mann. 1914. The ants of Haiti. Bull. Amer. Mus. Nat. Hist. 33(1):1–61.

Wilson, E.O. 1964. The true army ants of the Indo-Australian area (Hymenoptera: Formicidae: Dorylinae). Pacific Ins. 6(3):427–483.

———. 1985a. Ants of the Dominican amber (Hymenoptera: Formicidae), 1: Two new myrmicine genera and an aberrant *Pheidole*. Psyche 92(1):1–9.

———. 1985b. Ants of the Dominican amber (Hymenoptera: Formicidae), 2: The first fossil army ants. Psyche 92(1):11–16.

———. 1985c. Ants of the Dominican amber (Hymenoptera: Formicidae), 3: The subfamily Dolichoderinae. Psyche 92(1):17–37.

———. 1985d. Invasion and extinction in the West Indian ant fauna: Evidence from the Dominican amber. Science 229:265–267.

———. 1985e. Ants of the Dominican amber (Hymenoptera: Formicidae), 4: A giant ponerine in the genus *Paraponera*. Israel J. Ent. 19:197–200.

Wilson, E.O., and R.W. Taylor. 1967. The ants of Polynesia. Pacific Ins. Monogr. 14:1–109.

10 · Distribution Patterns and Biology of West Indian Sweat Bees (Hymenoptera: Halictidae)

George C. Eickwort

Sweat bees, members of the family Halictidae, are usually among the dominant members of bee faunas throughout the world, both in numbers of species and numbers of individuals. For instance, halictids represent 53–60% of the bee species in southern Brazil (Sakagami et al. 1967, Laroca et al. 1982), 18% of the bee species in Guanacaste, Costa Rica (Heithaus 1979), and 19% of the bee species in Carlinville, Illinois (Robertson 1929). Similar figures pertain to the West Indies. Halictidae comprise 32% of the bee species known from Jamaica (Raw 1985), 31% of the bee species recorded from Cuba by Alayo (1973), and 33% of the bees known from San Salvador, Bahamas (Elliott 1984). In numbers of individuals, Morse (1960) noted that halictids comprise 60–70% of bees captured in sweep samples in the northeastern United States, and Sakagami et al. (1967) found that halictids comprise 58% of the individual bees in southern Brazil.

The majority of halictid species excavate nests in the soil, frequently in partially bare and disturbed sites, although some nest in dead wood. Although many other soil-nesting bees are restricted to a narrow range of plants for pollen (i.e., they are oligolectic), the majority of halictids are opportunistic in their foraging, using a wide variety of pollen sources (i.e., they are polylectic). In short, halictids are often "weed bees," well adapted to living in close proximity to humans. Their lack of floral specialization has enabled members of the subfamily Halictinae to produce more than one generation per year, and thus to evolve a wide diversity of primitive societies (Michener 1974).

Most mainland species of Halictidae are widely distributed and are not restricted to localized food resources or specialized nest substrates. They are reasonably good colonizers, since a single inseminated female can establish a viable nest and theoretically become the foundress of a new population. On

231

Figure 10-1. A sweat bee, *Augochlora* sp. (Halictidae). (Photograph by Donald Specker.)

the other hand, they are not as strong fliers as are carpenter bees and euglossine bees, and are unlikely to make "voluntary" long-distance flights over water in order to locate new nest sites. They are also not as likely to be inadvertently rafted or transported by humans across water as are megachilid bees, carpenter bees, and other bees that nest in wood and natural cavities (Michener 1979). There is only one well-substantiated record of an accidental human-assisted dispersal of a sweat bee in North America—the establishment of the eastern species *Dialictus imitatus* in southern California (Michener and Wille 1961, Moure and Hurd 1987).

In a historical analysis of the insect fauna of the West Indies, sweat bees stand in contrast to other groups considered in this volume. Species of Halictidae should not be excluded from any large, mountainous island because of the lack of specific trophic or edaphic requirements, nor should a species be ecologically restricted to any one island. Halictids should be able occasionally to cross narrow water gaps and establish new populations, but regular long-distance migration between islands is unlikely. There are sufficient genera (13) to establish multiple phylogenies, enabling tests for concordant patterns of evolution; but the number of species is not large enough (30 in Cuba, 22 in Jamaica) to discourage the undertaking of taxonomic studies.

More importantly, sweat bees are of recent evolutionary origin. Apoidea probably first evolved in the Middle or Upper Cretaceous (Michener 1979). All genera of Halictidae except one in the West Indies also occur on the adjoining mainland. I hypothesize a mid-Tertiary or later origin of the West Indian halictid fauna, because it is unlikely that all of these genera attained their modern forms in the Mesozoic. This means that these bees reached the islands after the Greater Antilles separated from the mainland, which occurred at the latest in the late Cretaceous to Eocene (Sykes et al. 1982, Anderson and Schmidt 1983, Donnelly 1985). They presumably crossed water gaps to reach the islands, as also hypothesized by Michener (1979). A vicariant distribution of Antillean Halictidae would require that the Antilles originated much later than geologic evidence currently indicates (see Donnelly, chap. 2).

A goal of my studies is to revise the West Indian genera and to prepare phylogenetic analyses of the Greater Antillean taxa, using both morphological and electrophoretic data. This analysis will enable a comparison of the dispersal-based distribution of the Halictidae with the hypothesized vicariance-based distributions of older taxa of animals (Rosen 1975). The present paper will describe the distribution of the West Indian halictid genera and draw conclusions concerning their mainland sources.

Despite considerable interest in social behavior in the Halictidae, only one behavioral study has been conducted in the West Indies, by Raw (1975, 1984) on three species in Jamaica. Fieldwork in the Bahamas and Greater Antilles in 1985 has enabled me to describe aspects of nesting biology of additional species. Using these preliminary studies, I shall contrast the behavior and ecology of West Indian sweat bees with that of their mainland relatives.

Methods

My systematic studies on the West Indian Halictidae have been in progress for less than a year at this writing. This paper discusses the distribution of the genera and important species groups, concentrating on the Greater Antilles. This analysis is based on the literature and on examination of specimens in major insect collections, especially those of Cornell University, the U.S. National Museum of Natural History (Smithsonian Institution), the American Museum of Natural History, the Museum of Comparative Zoology at Harvard University, the Canadian National Collection, the British Museum (Natural History), the Cuban National Academy of Science, the Institute of Jamaica, the University of Puerto Rico, and collections by N. Elliott, A. Pauly, A. Raw, and M. Ivie. Knowledge of mainland taxa is based on my ongoing species-level studies of North American Halictinae and generic studies of Neotropical Halictinae (see Eickwort 1969).

Only one genus of Halictidae, *Agapostemon*, has been revised for the West

Indies (Roberts 1972). Species in other genera were determined by direct comparisons with type specimens or other specimens determined by the original authors. Seventy species names exist for West Indian Halictidae. I can associate names for all except three of these with specimens currently on loan to Cornell University. Numerous unnamed species are also available, but they do not represent new genera or subgenera and will not be named in this paper.

My assistants and I conducted field studies in 1985 to collect material for morphologic and electrophoretic analysis, to ascertain habitat and floral preferences, and especially to excavate nests. Nests were excavated following the techniques of Abrams and Eickwort (1980); contents were preserved in Dietrich's (Kahle's) fixative, and adults were dissected later at Cornell University. We visited San Salvador of the Bahama Islands (26 June–2 July; base of operations: College Center of the Finger Lakes Bahamian Field Station [CCFL]); Puerto Rico (1–14 May; bases of operations: University of Puerto Rico campus at Mayaguez and El Verde Field Station of the Center for Energy and Environmental Research, University of Puerto Rico); Dominican Republic (24–27 July; base of operations: Alcoa Exploration Company at Cabo Rojo); and Jamaica (31 July–15 August; bases of operations: Blue Mountain Field Station and Silver Sands at Duncans). Further fieldwork took place during 1986, and a detailed report on nesting biology will be published separately. Voucher specimens are deposited in the Cornell University Insect Collections, lot no. 995.

Results: Distribution Patterns

All three subfamilies of Halictidae—the Dufoureinae, Nomiinae, and Halictinae—occur in the Western Hemisphere (Michener 1979). Dufoureinae, the most primitive subfamily and the sister group of the rest of the Halictidae, is principally holarctic and African, with a relict genus in Chile. It is absent in the West Indies. Nomiinae is primarily a tropical and subtropical group of the Old World. Two groups of *Nomia* also occur in North America. They are most diverse in western North America and occur south into Mexico. One species is known from the West Indies.

The Halictinae is the largest and most cosmopolitan subfamily; its members are common components of nearly all bee faunas. The frequent behavior of licking human perspiration and other sources of salts has given them the colloquial name of "sweat bees." Two tribes occur in the Western Hemisphere. The Augochlorini is Neotropical in origin, with its most primitive members occurring in the South Temperate Zone (Eickwort 1969, Eickwort and Sakagami 1979). The ranges of three genera extend well into temperate North America. The Western Hemisphere Halictini has two components.

The first is a complex of genera centered about *Agapostemon*, which, like the Augochlorini, is Neotropical in origin with its most primitive members in the South Temperate Zone. The range of one genus extends well into temperate North America. The second component of the Halictini is cosmopolitan and presumably Old World in origin. In the Western Hemisphere, this component is most diverse at the generic level in North America, although there is considerable extension of these taxa into Central and South America and two genera have speciated extensively in the Neotropics. All components of the Halictinae are well represented in the West Indies.

There are three likely dispersal routes by which the Greater Antilles could have received sweat bees: (1) from Florida to Cuba and the Bahamas; (2) from Meso-America (Yucatán and Central America) to Cuba and Jamaica; and (3) from northern South America via Trinidad and the steppingstones of the Lesser Antilles. Similarly, there are two major dispersal routes by which the Lesser Antilles could have received bees: (1) from the Greater Antilles, and (2) from northern South America via Trinidad. The distribution of the halictid genera and principal species groups (Table 10-1) indicate the relative importance of these routes. Four of these genera (*Augochlora*, *Agapostemon*, *Dialictus*, and *Habralictellus*) contain the majority of the West Indian halictid species.

Origin in South America

Five genera or subgenera of Halictinae that occur in South America (including Trinidad) are known from the Lesser Antilles but not from the Greater Antilles. *Pseudaugochloropsis* is a genus of semisocial bees (Michener 1974), and a single species closely related to the widespread *P. graminea* has been collected in Grenada, St. Vincent, and Dominica. *Augochlora* (*Oxystoglossella*) is a soil-nesting subgenus of eusocial augochlorine bees (Eickwort and Eickwort 1972a); one species has been described from St. Vincent. *Neocorynura* is one of the largest genera of Augochlorini; its mainland species are solitary (Michener et al. 1966). One species has been collected on St. Vincent. *Habralictus* is a genus of Halictini related to *Agapostemon*, whose mainland species build solitary or communal nests (Michener et al. 1979); several species have been collected in Grenada, St. Vincent, and Dominica. *Microsphecodes* is a rarely collected genus of social parasites in the nests of other halictines; known mainland hosts are species of *Dialictus* and *Habralictus* (Eickwort and Eickwort 1972b, Michener et al. 1979). Three species have been described from Dominica and St. Vincent (Eickwort and Stage 1972).

These five genera or subgenera are Neotropical in origin and do not occur north of Central America (*Microsphecodes*), Mexico (*Neocorynura, Habralictus*), or southernmost Texas (*Pseudaugochloropsis, Oxystoglossella*).

Table 10-1. Distribution of halictid genera and important species groups in the West Indies and adjoining mainland

Taxon	Bahamas	Cuba	Jamaica	Hispaniola	Puerto Rico	Lesser Antilles	Florida	Meso-America	Trinidad & South America
Nomiinae									
Nomia (*Acunomia*)	X	X					X		
Halictinae: Halictini									
Halictus ligatus	X	X	X				X	X	X
Sphecodes	X	X	X	X	X	X	X	X	X
Microsphecodes		X	X	X		X		X	X
Dialictus									
gundlachi sp. group	X	X	X	X	X	X	X		X
mestrei sp. group		X	X	X					
longifrons sp. group		X	X	X			X		
ferreri sp. group		X	X	X	X		X		
parvus sp. group	X	X	X	X	X	?	X		
havanensis sp. group		X	X	X					
balophita sp. group		X	X	X			X		
sp. group A	X	X		X					
Habralictellus									
sp. group A	X	X	X	X	X	X			
sp. group B	X	X	X	X	X				
sp. group C	X	X		X					
Habralictus	X					X		X	X
Agapostemon									
poeyi sp. group	X	X	X	X	X		X	X	X
viridulus sp. group		X	X	X	X				
Halictinae: Augochlorini									
Augochlora (*Augochlora*)									
magnifica sp. group	X	X	X	X	X	X	X	X	X
"*praeclara*" sp. group	X	X	X	X	X				
species group A	X	X	X						
Augochlora (*Oxystoglossella*)				X					
Temnosoma		X	X					X	X
Pereirapis				?				X	X
Pseudaugochloropsis						X		X	X
Neocorynura						X		X	X

They are very distantly related to each other and represent independent invasions of the Lesser Antilles. The Antillean species are few and similar to mainland species, implying recent immigration.

Origin in North America

Two holarctic genera are represented by one species each in the Bahamas and the Greater Antilles but not in the Lesser Antilles, so dispersal from Florida or Meso-America is hypothesized. *Nomia* (*Acunomia*) *robinsoni* is an endemic species known from Eleuthera Island of the Bahamas and from Cuba, but not the other Greater Antilles (Ribble 1965). It is the only member of the Nomiinae present in the West Indies. One species of *Acunomia* (*nortoni*) occurs in Florida and a second (*howardi*) in Mexico, although not on the Yucatán peninsula.

Halictus (*Halictus*) *ligatus* has the widest range of any North American halictid bee, extending from Canada to Colombia, Venezuela, and Trinidad. It is an abundant species throughout the Bahamas, Cuba, and Jamaica, although it is absent from the other Greater Antilles. The island populations appear indistinguishable from mainland populations in morphology and nesting behavior, suggesting a recent dispersal from either Florida or Meso-America.

In contrast, *Dialictus* (often considered a subgenus of *Lasioglossum*) is holarctic in origin, (Michener 1979) but it has extensively radiated in South America, where endemic species groups occur. It is the most diverse genus of Halictidae throughout the Western Hemisphere, with considerable mixing in Central America of what appear to be originally North and South American species. It is also the most diverse halictid genus in the West Indies and contains the most individuals in most habitats.

Eight species groups of *Dialictus* occur on more than one of the Greater Antilles (Table 10-1), with little morphologic differentiation among the islands within each group. Cuba plus the Bahamas, Hispaniola, and Puerto Rico plus the Virgin Islands each contain species not found on the other islands. Most of the Greater Antillean species groups have no close relatives in the Lesser Antilles; only the *parvus* species group has a possible relative there.

Several of these species groups also occur in Florida. *Dialictus halophitus* is a halophilic coastal species (Graenicher 1930) that occurs from Louisiana to Florida to North Carolina (Moure and Hurd 1987). Floridian and Cuban specimens of *D. halophitus* are indistinguishable. Floridian specimens of *D. coreopsis* are conspecific with Cuban *D. longifrons*. *Dialictus coreopsis* occurs over a wide area of the eastern and central United States (Moure and Hurd 1987), and midwestern specimens are less similar to the West Indian *D. longifrons*. *Dialictus surianae*, a species known only from peninsular Florida

(Mitchell 1960), belongs to the *D. parvus* species group; and *D. flaveriae*, also known only from peninsular Florida (Mitchell 1960), belongs to the *D. gundlachi* species group. Other Greater Antillean species groups do not have members in the nearctic fauna. The brilliant green *Dialictus havanensis* species group in particular has no close relatives in Florida. My studies on *Dialictus* are still in progress, so I cannot yet report on Meso-American relatives on Antillean species. It is quite possible that some or all of the Greater Antillean species groups originated in Meso-America, spread through the islands, and then emigrated to Florida. Certainly the presence of several Cuban species in Florida suggests recent migration. The limited ranges of *D. surianae* and *D. flaveriae*, at least, suggest that they may be the results of emigration from the West Indies rather than in the opposite direction.

Neotropical Origin via Meso-America

Two (possibly three) Neotropical genera occur in the Greater Antilles but not the Lesser Antilles, suggesting dispersal from Meso-America. *Temnosoma* is a morphologically specialized genus of cleptoparasitic augochlorine bees whose species superficially more closely resemble chrysidid wasps than bees (Eickwort 1969). A species that appears identical with the Meso-American *T. smaragdinum* has been occasionally collected in Cuba and Jamaica.

Pereirapis is a small genus of very common eusocial augochlorine bees in Central and South America. Vachal (1911) named *Halictus cerasis* for a male bee, labeled from Haiti, which belongs in *Pereirapis* (pers. obs. of holotype in the Paris Museum). The genus has not otherwise been collected from the West Indies, and I did not see it in the Dominican Republic near the Haitian border. Until more specimens are located, I consider it doubtful that *Pereirapis* are present in the West Indies.

Agapostemon has been revised by Roberts (1972). Two species groups occur on more than one of the Greater Antilles, each quite distinct from the other and from mainland species groups. Roberts (1972) chose to name different species for the island forms in each species group for which he could detect consistent color differences; they could have easily been considered subspecies, although there is some geographic overlap in the Bahamas. He named the species groups by the names of the Cuban representatives, a practice I have followed for the other genera (Table 10-1). The *poeyi* species group is represented on all the Greater Antilles and many of the Bahamas, while the *viridulus* species group is absent from Puerto Rico and the Bahamas. Cuba and Hispaniola each possess unique species of *Agapostemon*. The distinctiveness of the species groups from mainland congeners and the tendency to form phenetically distinct local populations suggest a relatively long residence of *Agapostemon* in the Greater Antilles and Bahamas. Although the genus also occurs in North America, including Florida, I hypothesize that it reached the Antilles via Meso-America because the Antillean species are phenetically closer to Central American species than to *A. splendens*, the Floridian species (Roberts 1972).

Augochlora is represented in both the Greater and Lesser Antilles as well as in South and Central America, with one widespread temperate North American species. All Greater Antillean species are placed in the subgenus *Augochlora sensu stricto* (Eickwort 1969). Most of these species fall into three species groups (Table 10-1). As in *Agapostemon*, representatives of the *magnifica* species group have been given different names on different islands; the group occurs on all the Greater Antilles. A second species group corresponds to "*praeclara*" of Alayo (1973) (but is not the true *praeclara*, a rarely collected Cuban species), and is known from Cuba, Jamaica, and the Bahamas. Cuba and Jamaica each possess unique species. The Greater Antillean species groups of *Augochlora* are distinct from the Lesser Antillean species and mainland species. A Neotropical origin of the Greater Antillean species groups via Meso-America is proposed.

Uncertain Origin

Sphecodes, like *Dialictus*, is diverse throughout the Western Hemisphere. *Sphecodes* occurs sparsely in both the Greater and Lesser Antilles, and until mainland species groups are better analyzed, relationships of Antillean species cannot be ascertained. It is a genus of cleptoparasitic bees; one Jamaican species is known to parasitize *Dialictus* (Raw 1985). Three species are known from Cuba (Alayo 1973) and two from Jamaica.

Endemic Genus

One genus, *Habralictellus*, recently named by Moure and Hurd (1982), is endemic to the West Indies. Specimens are uncommon in collections. Three species groups occur on more than one of the Greater Antilles (Table 10-1). Species group A, collected near the coast, is very similar throughout the Greater Antilles and Bahamas, while the high-altitude species group B has phenetically distinguishable populations on each of the Greater Antilles. Unique species are known from Cuba and from Hispaniola, but not Jamaica or Puerto Rico. A different set of species groups occurs in the Lesser Antilles (Dominica, St. Vincent, and Montserrat). *Habralictellus* belongs to the *Lasioglossum* complex of genera and is most closely related to *Dialictus*. I hypothesize that the stem species of *Habralictellus* evolved from *Dialictus* soon after the latter genus reached the Greater Antilles. *Habralictellus* then speciated and the daughter species spread through the islands, with one or more lineages extending into the Lesser Antilles.

Absent Genera

Of equal interest are those halictid genera present on the adjacent mainland that are absent on the West Indies. In Trinidad and Meso-America, species of *Augochloropsis, Pereirapis*, and the nocturnal *Megalopta* are abundant augochlorine bees that would appear to be well adapted to weedy and forested habitats in the West Indies. In Florida, species of *Halictus* (*Seladonia*), *Evylaeus*, and *Augochlorella* would also appear to be well adapted to the adjacent West Indies. Other halictid genera and subgenera present on the adjoining mainland but absent in the West Indies are *Lasioglossum sensu stricto* and *Nomia* (subgenera *Dieunomia, Epinomia*, and *Curvinomia*) (nearctic origin) and *Caenohalictus, Megommation* (*Megaloptina*), and *Caenaugochlora* (Neotropical origin).

Results: Ecology

Our field observations were directed toward investigating whether West Indian halictid bees play the same ecological roles as do their mainland counterparts. A major family of ground-nesting bees, the Andrenidae, is almost absent from the West Indies (Michener 1979) (two undescribed species of Panurginae occur in Hispaniola) as are numerous genera in other families. The stingless bees (Meliponinae), a major group of Neotropical social bees, are also currently absent from the Greater Antilles (except for the apparently human-introduced *Melipona fulvipes* in Jamaica and Cuba; Raw 1985, Alayo 1973) as are the eusocial bumble bees and the orchid bees (except for *Euglossa jamaicensis* in Jamaica). Consequently, the Halictidae are the only native eusocial bees in the Greater Antilles. The family represents about half the species of soil-nesting bees there (24 of 49 species, excluding Megachilidae, in Cuba; Alayo 1973). We predicted that halictid bees might therefore occur in a greater diversity of habitats than they do on the mainland, perhaps visiting a wider range of plants. Conversely, halictids might have evolved into floral specialists (oligoleges), a role played especially by the Andrenidae in arid areas of the mainland. At the least we expected them to be numerous in most habitats on the islands.

Our observations did not support these predictions. Halictid bees were surprisingly sparse on flowers in many habitats, especially in the highlands. Intensive searching produced few or no sweat bees in many locations that appeared ecologically identical to those that support an abundant and diverse halictid fauna in Florida, Costa Rica, and Brazil. Sweat bees were abundant in some (but not all) highly disturbed lowland locations, especially just behind the coast, and in scattered upland locations. Such habitats reminded us of disturbed locations on the mainland where halictids are also common. Raw (1985) and Michener (pers. comm.) noted that bees in general are more diverse and abundant in coastal and midelevation areas than in forested uplands in Jamaica. We also observed this general pattern in Puerto Rico. In most locations Halictidae were the most common Apoidea that we collected, along with *Exomalopsis* (Anthophoridae) and the introduced honey bee, *Apis mellifera*. Raw (1985) also noted that Halictidae are the predominant small bees in Jamaica.

The flowers that halictids visited in our studies were usually the same weedy plants they frequent in comparable mainland locations. There was no evidence that sweat bees have expanded their host ranges to include plants that they would not visit on the mainland. Nor have they evolved toward oligolecty, although in any one locality individual halictid species (especially of *Agapostemon, Augochlora, Habralictellus*, and *Halictus*) were frequently concentrated on one pollen source, a phenomenon that is also frequent on the mainland.

Halictus ligatus exhibits remarkable ecological plasticity in its extended mainland range (Michener and Bennett 1977). In the Bahamas, Jamaica and Cuba it occurs only in the lowlands, where it nests in highly disturbed, partially vegetated areas similar to its mainland nest sites (Raw 1984, Alayo 1973). In the Bahamas and Jamaica, we corroborated Raw's finding that the composite weed *Bidens pilosa* is the principal pollen source. This is the same plant preferred by *H. ligatus* in Florida and Costa Rica (pers. obs. and Packer and Knerer 1986) and in Trinidad (Michener and Bennett 1977). Ecologically, the West Indian and the tropical mainland populations of *H. ligatus* appear to be identical (Raw 1984).

The various species of *Dialictus* are small, dull-green bees that are typically the most abundant bees in disturbed, weedy locations throughout the Western Hemisphere. Their inconspicuous appearance and preference for low-growing weeds with small flowers (e.g., *Euphorbia*) often leads a casual collector to underestimate their abundance. We found them in similar disturbed habitats in the West Indies, visiting the same plants as would be anticipated from mainland studies.

Species of *Agapostemon* are generally among the most conspicuous of bees, with brilliant green or bluish heads and thoraces and concolorous or black-and-yellow abdomens. As such they are rarely ignored by general collectors. On the mainland they are best represented in open areas and are especially numerous in beach and desert locations. Although *Agapostemon poeyi* and *viridulus* are described as being abundant throughout Cuba by Alayo (1973), we observed widely scattered populations of *Agapostemon*. At all our locations, females collected pollen early in the morning (0700) and often ceased by 0900, so we suspect that the bees are often missed by general collectors. Males continued to patrol vegetation later in the day in some sites. Early-morning foraging has also been noted for mainland species of *Agapostemon* (Roberts 1969), although the bees typically continue foraging later in the day than we observed in the West Indies.

Augochlora (*Augochlora*) (Fig. 10-1) is a subgenus of wood-nesting bees on the mainland (Eickwort and Eickwort 1973), and we predicted that the same dead-wood substrate would be used for nests in the West Indies, and that the species would consequently be most common in forested areas. The brilliant-green bees are conspicuous in flight and are rarely ignored by general collectors. We did not find any species to be common in the habitats we visited, although Alayo (1973) recorded *A. magnifica* and *A. "praeclara"* as being abundant throughout Cuba, and Raw (1985) recorded *A. regina* (*magnifica* species group) as common on a mangrove spit in Jamaica. We did indeed most frequently collect *Augochlora* at the edges of forests. We did not succeed in locating nests.

Habralictellus is the endemic halictid genus in the West Indies, and therefore the one that we predicted would be most likely to show adaptations for island living. We thought that the genus might be restricted to relatively undisturbed habitats, perhaps to the highlands, and that its species might exhibit specializations for endemic flowers. The species of *Habralictellus* are not common in museum collections; Alayo (1973) collected each species (his *Lasioglossum* spp. A to H) in one or only a few locations in Cuba. Despite diligent searching for these small, brilliantly colored bees, we rarely encountered them. Representatives of species group B were collected sparsely at high elevations, in a moist forest at 1000 m in Puerto Rico and at the edge of a forest at a slightly lower elevation in the Sierra de Bahoruco in the Dominican Republic. Representatives of species group A were collected on a flowering palm in the lawn of the College Center of the Finger Lakes field station in San Salvador, on a mowed lawn at Luquillo Beach in Puerto Rico, and on weeds at the base of a water tank in thorn scrub at Cabo Rojo, Dominican Republic. In San Salvador and the Dominican Republic, *Habralictellus* were quite abundant on the highly localized pollen sources, but we did not find them elsewhere in the surrounding areas. We do not believe that these bees are oligolectic; they simply appeared to be concentrating on the most prolific pollen and nectar sources in these areas. We were unable to locate nests, and I cannot generalize as to the adaptations of species of *Habralictellus* or even predict where they will next be collected; certainly they are not restricted to undisturbed habitats.

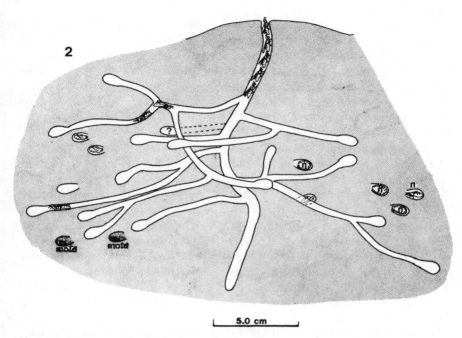

Figure 10-2. Communal nest of *Agapostemon kohliellus*, Cabo Rojo, Dominican Republic.

Results: Nesting Biology

The nesting biology of mainland species of all West Indian genera of Halictidae (except the endemic *Habralictellus*) is known (reviewed in Michener 1974, Eickwort and Sakagami 1979). This allows comparison of the West Indian species with their close mainland relatives to ascertain whether there are significant adaptations in nest structure and social behavior correlated with island existence.

Agapostemon kohliellus, apparently endemic to Hispaniola, nested in a dense aggregation in the Alcoa residential area in Cabo Rojo, Dominican Republic. Nests occurred in human-deposited volcanic silt overlying coral rock, directly adjoining a residence and shaded by cultivated citrus trees. The site was watered during the dry season by gardeners.

Nests consisted of rather short vertical main burrows leading to an extensive system of long lateral burrows, each leading to one or more open cells (Fig. 10-2). After provisioning and oviposition, the laterals were filled with soil, so nests excavated in late afternoon consisted only of main burrows. The nests were evidently enlarged each evening by new excavation of lateral burrows and cells. The nests ($n = 10$) contained from 1 to 10 bees. All females were inseminated and had developing oocytes in their ovarioles ($n = 21$); the bees were therefore all potentially reproductive and the colonies were communal societies, as defined by Michener (1974). Such communal associations are characteristic of many (but not all) Neotropical and Nearctic species of *Agapostemon* (Eickwort and Eickwort 1969, Abrams and Eickwort 1980, Eickwort 1981). The nest architecture of *A. kohliellus* is also characteristic of the genus.

The cleptoparasitic ("cuckoo") bee *Nomada* (*Micronomada*) *krugii* (Anthophoridae: Nomadinae) was abundant in the *A. kohliellus* nest site. Female cuckoo bees frequently investigated nest entrances and were observed to enter and stay within nests. Excavated cells contained *Nomada* immatures. *Nomada krugii* belongs to the same subgenus as *N. formula*, which parasitizes Nearctic *Agapostemon* (Eickwort and Abrams 1980). Abrams and Eickwort (1981) showed that guards of *Agapostemon virescens* effectively protect nests from attack by *Nomada articulata*, and hypothesized that cooperative nest guarding is an important advantage of communal nesting, but such guarding (if it occurs) was not effective in protecting nests of *A. kohliellus* from attack by *N. krugii*. No other parasites were observed to enter nests.

Dialictus parvus nested in loose aggregations in an automobile parking area on the CCFL field station grounds in San Salvador and in an abandoned homesite nearby. Both sites were coral sand and pulverized rock, with sparse low vegetation, and in full sun. *Dialictus gemmatus*, a member of the *parvus* species group, nested in a loose aggregation in the upper beach area of Silver Sands, Jamaica. Nests were in a sparsely grassy area used for automobile parking. *Dialictus gundlachi* and *D. ferreri* nested in the grassy upper beach area of Luquillo Beach, Puerto Rico. Each species nested in an aggregation under a separate palm tree, in bare sandy soil.

The nests of these species of *Dialictus* were similar in their basic architecture and contents. Each consisted of a single vertical burrow with laterals leading to individual cells (Figs. 10-3–10-5). Nests contained only a few cells (maxima of 6 to 10, medians = 4 for each species) and few bees (maxima of 4 or 5, medians = 2 or 3 for each species). In five nests of *D. parvus* and *gemmatus*, the largest bee was the only one that was inseminated and had developed ovaries; the other, smaller bees were unmated and had undeveloped ovaries and were assumed to be workers. However, one other nest had two mated bees with undeveloped ovaries (one bee escaped). Of three multifemale nests of *D. gundlachi*, two contained one reproductive and one worker bee each, but the third contained four inseminated, reproductive females and no worker bees. Three multifemale nests of *D. ferreri* contained only inseminated, reproductive females. I conclude that nests of *D. parvus* species group and *D. gundlachi* are typically eusocial or semisocial, but

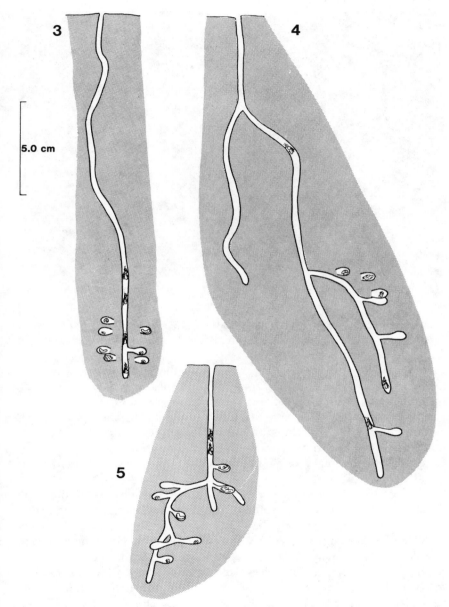

Figure 10-3 to 10-5. Nests of *Dialictus*. 3, *D. parvus*, San Salvador, Bahamas; 4, *D. ferreri*, Luquillo Beach, Puerto Rico; 5, *D. gundlachi*, Luquillo Beach, Puerto Rico.

Figure 10-6. Nest sites of *Dialictus ferreri* and *gundlachi*, Luquillo Beach, Puerto Rico.

occasionally can be communal, which is the usual state for nests of *D. ferreri*. Castes are weakly defined, the queens being only slightly larger than the workers in most nests.

No parasites were observed around or in the nests of any species of *Dialictus*, although Raw (1985) recorded the cleptoparasite, *Sphecodes* sp., in the nests of *Dialictus liguanensis* (*gundlachi* species group) in Jamaica. According to Raw (pers. comm.), *D. liguanensis*, like *D. gundlachi*, is social. The nesting biology of all the West Indian *Dialictus* closely resembles that of many mainland Neotropical *Dialictus* (Michener et al. 1979), which also have surprisingly small colonies, in terms of numbers of cells and adult females.

Nests of *Halictus ligatus* were located in San Salvador, in the homesite location of *D. parvus*, and also in Jamaica, in red volcanic silt at the edge of a farm road. In both locations, the nests were widely scattered and in fully insolated soil near the edges of low plants. The nests had large tumuli, were guarded, and had many occupants. The soil was too hard for us to excavate the deep Bahamian nests to cell level, and we did not have the opportunity to excavate the Jamaican nests. Raw (1975, 1984) excavated three deep nests in Jamaica, containing 10, 10, and 13 female bees, the largest nest with 57 cells. He reported that all bees had developed ovaries. The nest architecture and social structure appear identical to those of Trinidadian *H. ligatus* examined by Michener and Bennett (1977), with poorly defined castes, and different from those of North American populations, where castes are better defined and workers are typically the daughters of the queens (Michener and Bennett 1977; Packer and Knerer 1986).

Discussion and Conclusions

The predominant direction of migration of halictid bees into the Greater Antilles has been from Meso-America to Cuba and/or Jamaica. Migration into the Greater Antilles from South America via Trinidad and the Lesser Antillean steppingstones has not played a significant role. The Lesser Antilles themselves have received some recent immigrants from South America, little differentiated from their mainland relatives. The Lesser Antillean fauna is distinctly different at the species-group level from the Greater Antillean fauna, with no species in common yet known. This pattern of faunation of the Greater and Lesser Antilles is also the predominant one for butterflies (Brown and Heineman 1972, Scott 1972, Brown 1978), birds (Bond 1978), and bats (Baker and Genoways 1978).

The major genera of Halictidae in the Greater Antilles, *Augochlora*, *Agapostemon*, and *Dialictus*, contain species groups that are morphologically distinct from their mainland Neotropical congeners. This implies a long period of separation of the West Indian populations. The species groups within each genus are also quite distinct from each other, implying either a separate migration of the ancestor of each species group into the Antilles or an extended period of species-level evolution in the islands. The latter hypothesis appears necessary for *Habralictellus*, which is represented by at least 6 species groups or unique species in the Greater Antilles and other groups in the Lesser Antilles, but does not occur on the mainland. Certainly each of the genera has undergone little speciation in the Antilles in comparison to the bee genus *Hylaeus* (*Neoprosopis*) in the Hawaiian Islands, in which over 50 species evolved from a single ancestor (Zimmerman 1970).

The actual source of the initial migrants poses an interesting problem for the principally Neotropical genera *Augochlora* and *Agapostemon*. If they migrated to the Greater Antilles from Meso-America (e.g., from the southern tip of North America) during the Tertiary separation of North and South America, either they had already evolved as genera in South America and dispersed to North America before the continents split apart 70 Ma, or they were able to disperse from South to North America during the Tertiary. The latter hypothesis is more credible. The fact that *Agapostemon* has four North American species groups, distinct from the South American species groups (Roberts 1972), adds credence to the hypothesis that that genus reached North America long before the Central American land bridge formed in the Pliocene. Also, today's Central American *Agapostemon* fauna is largely a mixture of North and South American species groups (Roberts 1972), implying that little higher-level evolution has taken place in Meso-America since the Pliocene land connection formed about 3 Ma.

Each of the major Greater Antillean halictid genera further exhibits species groups that are usually represented on more than one island, by populations that are morphologically indistinguishable or separated by only minor characteristics. This implies that at some time in the recent past the major islands were sufficiently close to each other that dispersal was readily undertaken, perhaps during low-water periods in glacial maxima of the Pleistocene (Pregill and Olson 1981).

However, not all species groups in the major halictid genera are present on all the Greater Antilles today (Table 10-1). Considering only those species groups that have been collected in reasonable numbers and from more than one island, Cuba has 17 groups represented, Jamaica 15, Hispaniola 12, and Puerto Rico 7. Each Greater Antillean island has unique species that have been collected in reasonable numbers and are so far known only from that island: Cuba has at least 5 (*Nomia robinsoni*, *Dialictus magdalena*, and 3 species of *Habralictellus*), Hispaniola at least 4 (*Agapostemon kohliellus*, 2 species of *Dialictus*, and 1 species of *Habralictellus*), Jamaica 1 (*Augochlora decorata*), and Puerto Rico 1 (a species of *Dialictus*). While additional collecting will probably expand the ranges of these species groups, especially in Hispaniola, I find it difficult to believe that all of them actually occur on all the major islands. The reduced number of species on Puerto Rico, and the few endemic species groups there, parallels the pattern seen in other animals (Brown and Heineman 1972, Scott 1972, Terborgh 1973, Thompson 1981, Briggs 1984).

Differential extinction of the species groups of halictids may have played a role in determining their current distribution in the Greater Antilles. During the Pleistocene, land areas may have been much reduced during interglacial periods, and forested areas may have been greatly reducing during these arid intervals (Pregill and Olson 1981). *Trigona*, the principal mainland genus of stingless bees and a typical forest inhabitant, occurs as a fossil (*Trigona dominicana*) in the lower Oligocene-middle Miocene Dominican amber (Wille and Chandler 1964, Michener 1982), but it is absent from the Greater Antilles today, obviously the victim of extinction. Wilson (1985) records 12 ant genera and subgenera present in the Dominican amber that are not present in the Greater Antilles today.

On the other hand, *Halictus*, *Temnosoma*, and four other species groups of *Dialictus* are represented by species that also occur on the mainland. These would appear to be the results of more recent dispersal (Pleistocene or Recent) to (or from) the West Indies than that of *Augochlora*, *Agapostemon*, and other species groups of *Dialictus*, such as the *havanensis* group. The presence of four species groups of *Dialictus* in peninsular Florida cannot be taken as evidence that each of these groups invaded the Greater Antilles from mainland North America. Migration in the opposite direction, from Cuba or the Bahamas to Florida in recent (perhaps historical) times, is equally possible, especially for the *D. parvus* and *D. gundlachi* species groups, whose mainland representatives are known only from peninsular Florida and are somewhat distinct from other United States species. Phylogenetic analysis of North American species of *Dialictus*, which I am currently undertaking, should resolve the direction of migration. Emigration from the Greater Antilles into Florida is hypothesized for other insect groups (e.g., Scott 1972, Liebherr 1985).

The restriction of such commonly collected, lowland species groups as *Halictus ligatus*, *Agapostemon viridulus*, *Augochlora "praeclara," Dialictus mestrei*, *D. longifrons*, and *D. halophitus* to certain of the Greater Antilles, and the absence of all from the Lesser Antilles (Table 10-1), is evidence that dispersal among the islands is not readily accomplished by sweat bees today. For instance, *H. ligatus* is a highly adaptable "weed bee" and conditions would seem appropriate for it on any island, even the smallest ones. *H. ligatus* has been able to disperse to various Bahamian islands, so its absence from Hispaniola and Puerto Rico is even more surprising. Moreover, important Neotropical and Nearctic genera of Halictidae did not succeed in becoming established in either the Greater or Lesser Antilles, although their mainland ecology suggests that they should be successful on the islands.

Ecologically, the species of Halictidae on the Greater Antilles display the attributes that are predominant in their closest tropical and subtropical mainland relatives. There is no indication that they have diversified into new life styles to take advantage of the near absence of such significant mainland groups as the floral-specialist Andrenidae or the eusocial stingless bees and bumble bees. They stand in contrast to *Hylaeus* (*Neoprosopis*) of the Hawaiian Islands, which evolved a diversity of habits, including cleptoparasitism, from a single ancestor (Linsley 1966, Zimmermann 1970).

Communal *Agapostemon* and eusocial or semisocial *Halictus* and *Dialictus* are indistinguishable in their social structure from their tropical mainland congeners. The environmental factors that select for these social structures are obscure. The small societies of *Dialictus* are similar among the three species groups studied on three islands over a three-month span. Why such societies should be favored rather than the large colonies often developed by North American eusocial species is not obvious. Similar small societies occur in most mainland Neotropical *Dialictus*; the common selective factors are likely to be climatic rather biotic (that is, escape from cleptoparasites or predators).

As Raw (1977, 1985) has noted for Jamaica, primitively eusocial and parasocial species (Halictidae and *Exomalopsis*) predominate among the small bees in lowland and disturbed habitats in the Greater Antilles. In Puerto Rico, all of the commonly collected species of *Dialictus*, the predominant small bees on the island along with the parasocial *Exomalopsis*, are social; only the much less common *Agapostemon viequesensis* (*poeyi* species group) and *Augochlora buscki* (*magnifica* species group) are likely to be solitary (the status of *Habralictellus* is unknown). The ecological factors that select for this predominance of primitively social ground-nesting bees are unknown. For a behaviorist, the West Indies provide an excellent natural laboratory in which to study the biology of primitively social sweat bees in a simplified ecosystem.

Acknowledgments

I thank Anthony Raw, Thomas Farr, Ronald McGinley, Jerome Rozen and Marjorie Favreau, Alain Pauly, Nancy Elliott, James Carpenter, and especially Pastor Alayo D. for the loan of specimens; Dr. Elliott and Dr. Raw for advice and unpublished manuscripts on biology of Caribbean bees; and Donald Gerace (San Salvador), Dr. Farr, Brian Freeman, Eric Garraway, David Smith (Jamaica), Rafael Reyes, Victor García, Ramón Caceres, Renato Rimoli (Dominican Republic), and James Ackerman, Julio Micheli, Flavio Padovani, J. A. Ramos, and especially Stuart Ramos (Puerto Rico) for much assistance during our visits to the islands. The success of our fieldwork is largely due to the untiring and uncomplaining efforts of Byron Alexander, Mary Eickwort, and most especially Anne Rosenthal. F. Robert Wesley ably assisted in laboratory and curatorial work at Cornell. I am grateful to James Liebherr for the invitation to contribute to the E.S.A. symposium and this volume, and for his advice on this research. Helpful comments on the manuscript were provided by Mr. Alexander, Drs. Raw and Liebherr, Charles Michener, Radclyffe Roberts, Stephen Nichols, Janis Dickinson, Monica Raveret-Richter, and Penelope Kukuk. This research was supported by National Science Foundation Grant No. BSR–8413229.

References

Abrams, J., and G.C. Eickwort. 1980. Biology of the communal sweat bee *Agapostemon virescens* (Hymenoptera: Halictidae) in New York State. Search: Agriculture (Cornell Univ. Agr. Exp. Sta.) 1980(1):20pp.

———. 1981. Nest switching and guarding by the communal bee *Agapostemon virescens* (Hymenoptera, Halictidae). Insectes Sociaux 28:105–116.

Alayo D., P. 1973. Catálogo de los himenópteros de Cuba. Insituto Cubano del Libro, Havana. 218 pp.

Anderson, T.H., and V.A. Schmidt. 1983. The evolution of Middle America and the Gulf of Mexico-Caribbean Sea region during Mesozoic time. Geol. Soc. Amer. Bull. 94:941–966.

Baker, R.J., and H.H. Genoways. 1978. Zoogeography of Antillean bats. *In* F.B. Gill, ed., Zoogeography in the Caribbean, pp. 53–97. Acad. Natur. Sci. Philadelphia, Spec. Pub. No. 13.

Bond, J. 1978. Derivations and continental affinities of Antillean birds. *In* F.B. Gill, ed., Zoogeography in the Caribbean, pp. 119–128. Acad. Natur. Sci. Philadelphia, Spec. Pub. No. 13.

Briggs, J.C. 1984. Freshwater fishes and biogeography of Central America and the Antilles. Syst. Zool. 33:428–435.

Brown, F.M. 1978. The origins of the West Indian butterfly fauna. *In* F.B. Gill, ed., Zoogeography in the Caribbean, pp. 5–30. Acad. Natur. Sci. Philadelphia, Spec. Pub. No. 13.

Brown, F.M., and B. Heineman. 1972. Jamaica and Its Butterflies. E.W. Classey, London. 478 pp.

Donnelly, T.W. 1985. Mesozoic and Cenozoic plate evolution of the Caribbean region. *In* F.G. Stehli and S.D. Webb, eds., The Great American Biotic Interchange, pp. 89–121. Plenum Press, New York.

Eickwort, G.C. 1969. A comparative morphological study and generic revision of the augochlorine bees (Hymenoptera: Halictidae). Univ. Kansas Sci. Bull. 48:325–524.

———. 1981. Aspects of the nesting biology of five nearctic species of *Agapostemon* (Hymenoptera: Halictidae). J. Kansas Entomol. Soc. 54:337–351.

Eickwort, G.C., and J. Abrams. 1980. Parasitism of sweat bees in the genus *Agapostemon* by cuckoo bees in the genus *Nomada* (Hymenoptera: Halictidae, Anthophoridae). Pan-Pac. Entomol. 56:144–152.

Eickwort, G.C., and K.R. Eickwort. 1969. Aspects of the biology of Costa Rican halictine bees, I: *Agapostemon nasutus* (Hymenoptera: Halictidae). J. Kansas Entomol. Soc. 42:421–452.

———. 1972a. Aspects of the biology of Costa Rican halictine bees, IV: *Augochlora* (*Oxystoglossella*) (Hymenoptera: Halictidae). J. Kansas Entomol. Soc. 45:18–45.

———. 1972b. Aspects of the biology of Costa Rican halictine bees, III: *Sphecodes kathleenae*, a social cleptoparasite of *Dialictus umbripennis* (Hymenoptera: Halictidae). J. Kansas Entomol. Soc. 45:529–541.

———. 1973. Notes on the nests of three wood-dwelling species of *Augochlora* from Costa Rica (Hymenoptera: Halictidae). J. Kansas Entomol. Soc. 46:17–22.

Eickwort, G.C., and S.F. Sakagami. 1979. A classification of nest architecture of bees in the tribe Augochlorini (Hymenoptera: Halictidae; Halictinae), with description of a Brazilian nest of *Rhinocorynura inflaticeps*. Biotropica 11:28–37.

Eickwort, G.C., and G.I. Stage. 1972. A new subgenus of Neotropical *Sphecodes* cleptoparasitic upon *Dialictus* (Hymenoptera: Halictidae, Halictinae). J. Kansas Entomol. Soc. 45:500–515.

Elliott, N.B. 1984. Field guide to the insects of San Salvador Island, Bahamas. College Center of the Finger Lakes Bahamian Field Station, San Salvador, Bahamas. 33 pp.

Graenicher, S. 1930. Bee-fauna and vegetation of the Miami region of Florida. Ann. Entomol. Soc. Amer. 23:153–174.

Heithaus, E.R. 1979. Community structure of Neotropical flower visiting bees and wasps: Diversity and phenology. Ecology 60:190–202.

Laroca, S., J.R. Cure, and C. de Bortoli. 1982. A associação de abelhas silvestres

(Hymenoptera: Apoidea) de uma área restrita no interior da cidade de Curitaba (Brasil): Uma abordagem biocenótica. Dusenia 13:93–117.

Liebherr, J.K. 1985. Univariate analysis of multivariate data to evaluate circular overlap in the *Agonum decorum* complex (Coleoptera: Carabidae). Ann. Entomol. Soc. Amer. 78:790–798.

Linsley, E.G. 1966. Pollinating insects of the Galápagos Islands. *In* R.I. Bowman, ed., The Galápagos, pp. 225–232. Proceedings of the Symposia of the Galápagos International Scientific Project. University of California Press, Berkeley.

Michener, C.D. 1974. The Social Behavior of the Bees. Harvard University Press, Cambridge. 404 pp.

———. 1979. Biogeography of the bees. Ann. Missouri Bot. Gard. 66:277–347.

———. 1982. A new interpretation of fossil social bees from the Dominican Republic. Sociobiology 7:37–45.

Michener, C.D., and F.D. Bennett. 1977. Geographical variation in nesting biology and social organization of *Halictus ligatus*. Univ. Kansas Sci. Bull. 51:233–260.

Michener, C.D., and A. Wille. 1961. The bionomics of a primitively social bee, *Lasioglossum inconspicuum*. Univ. Kansas Sci. Bull. 42:1123–1202.

Michener, C.D., M.D. Breed, and W.J. Bell. 1979. Seasonal cycles, nests, and social behavior of some Colombian halictine bees (Hymenoptera; Apoidea). Rev. Biol. Trop. 27:13–34.

Michener, C.D., W.B. Kerfoot, and W. Ramírez B. 1966. Nests of *Neocorynura* in Costa Rica (Hymenoptera: Halictidae). J. Kansas Entomol. Soc. 39:245–258.

Mitchell, T.B. 1960. Bees of the eastern United States, vol. 1. North Carolina Agr. Exp. Sta. Tech. Bull. 141. 538 pp.

Morse, R.A. 1960. The abundance of wild bees (Apoidea) in the northeastern United States. J. Econ. Entomol. 53:679–680.

Moure, J.S., and P.D. Hurd, Jr. 1982. On two new groups of neotropical halictine bees (Hymenoptera, Apoidea). Dusenia 13:46.

———. 1987. An annotated catalog of the halictid bees of the Western Hemisphere (Hymenoptera: Halictidae). Smithsonian Inst. Press, Washington, D.C. 405 pp.

Packer, L., and G. Knerer. 1986. The biology of a subtropical population of *Halictus ligatus* Say (Hymenoptera: Halictidae), 1: Phenology and social organisation. Behav. Ecol. Sociobiol. 18:363–375.

Pregill, G.K., and S.L. Olson. 1981. Zoogeography of the West Indian vertebrates in relation to Pleistocene climatic cycles. Ann. Rev. Ecol. Syst. 12:75–98.

Raw, A. 1975. Studies on tropical and temperate bees. Ph.D. diss., University of the West Indies, Jamaica.

———. 1977. The biology of two *Exomalopsis* species (Hymenoptera: Anthophoridae) with remarks on sociality in bees. Rev. Biol. Trop. 25:1–11.

———. 1984. The nesting biology of nine species of Jamaican bees (Hymenoptera). Rev. Brasil. Entomol. 28:497–506.

———. 1985. The ecology of Jamaican bees (Hymenoptera). Rev. Brasil. Entomol. 29:1–16.

Ribble, D.W. 1965. A revision of the banded subgenera of *Nomia* in America (Hymenoptera: Halictidae). Univ. Kansas Sci. Bull. 45:277–359.

Roberts, R.B. 1969. Biology of the bee genus *Agapostemon* (Hymenoptera: Halictidae). Univ. Kansas Sci. Bull. 48:689–719.

———. 1972. Revision of the bee genus *Agapostemon* (Hymenoptera: Halictidae). Univ. Kansas Sci. Bull. 49:437–590.

Robertson, C. 1929. Flowers and Insects. Science Press Printing Co., Lancaster, Penn. 221 pp.

Rosen, D.E. 1975. A vicariance model of Caribbean biogeography. Syst. Zool. 24:431–464.

Sakagami, S.F., S. Laroca, and J.S. Moure. 1967. Wild bee biocoenotics in São Jose dos Pinhais (PR), South Brazil. Preliminary report. J. Fac. Sci., Hokkaido Univ., ser. 6 Zool. 16:253–291.

Scott, J.A. 1972. Biogeography of Antillean butterflies. Biotropica 4:32–45.

Sykes, L.R., W.R. McCann, and A.L. Kafka. 1982. Motion of Caribbean plate during last 7 million years and implications for earlier Cenozoic movements. J. Geophys. Res. 87 (no. B 13):10, 656–10, 676.

Terborgh, J. 1973. Chance, habitat, and dispersal in the distribution of birds in the West Indies. Evolution 27:338–349.

Thompson, F.C. 1981. The flower flies of the West Indies (Diptera: Syrphidae). Mem. Entomol. Soc. Washington 9. 200 pp.

Vachal, J. 1911. Etude sur les *Halictus* d'Amérique (Hym.). Misc. Entomol. Narbonne 19:9–24, 41–56, 107–116.

Wille, A., and L. Chandler. 1964. A new stingless bee from the Tertiary amber of the Dominican Republic (Hymenoptera; Meliponini). Rev. Biol. Trop. 12:187–195.

Wilson, E.O. 1985. Invasion and extinction in the West Indian ant fauna: Evidence from the Dominican amber. Science 229:265–267.

Zimmerman, E.C. 1970. Adaptive radiation in Hawaii with special reference to insects. Biotropica 2:32–38.

11 · Fossils, Phenetics, and Phylogenetics: Inferring the Historical Dynamics of Biogeographic Distributions

Edward F. Connor

The goal of historical biogeography is to describe and understand how the present geographical distributions of plants and animals came to be. To do so, one must know the present distribution of taxa and have some means of inferring their historical dynamics. The former of these two tasks can be achieved by exhaustive collecting and identification. How to perform the latter has been and still is the subject of intense debate.

Geographic distributions change because of the dispersal of organisms, the dispersal or division of geographic regions (vicariance), adaptive radiation, and extinction. However, recognizing the mark of each of these forces in specific geographic regions is a devilishly difficult proposition. This is largely because we rely on our imperfect knowledge of the current distribution of species and higher taxa, rather than on direct evidence of the historical dynamics of geographic distributions.

Our desire to discern the relative contributions of these forces in shaping geographic distributions is by no means new. Neither are the polarized views that either dispersal of organisms or dispersal of continents accounts for most of the current and historical distributions of species. What is new in this discussion is a more critical view of the role of fossil evidence and the "phenetic" assessment of the relationship between geographic regions. More importantly, the realization that the history of geographic distributions must to some extent mirror the history of life (Croizat et al. 1974, Rosen 1978) may provide a sound basis for inferring patterns in the development of geographic distributions, if not their underlying causes.

Much of the current debate concerning dispersal and vicariance has been played out on the stage of the Caribbean (see Liebherr, chap. 1). This region is particularly suited to serve as the "model" arena for developing and testing techniques of historical biogeographic inference, primarily because it is a region of great geological and biological complexity. Therefore, approaches that succeed here are bound to be useful elsewhere.

In the contributions to this book, we have seen general agreement that much remains to be done to develop a complete picture of geographic distribution, entomological and otherwise, in the Caribbean region. Neither the islands nor the adjacent mainlands have been sufficiently collected to provide a precise picture of regional distributions. We also have seen general disagreement concerning the types of evidence that can provide a firm basis for inferring the historical biogeographic dynamics of this region. The contributions to this volume epitomize the diverse approaches currently used by biogeographers, ranging from analyses of fossils to the biogeographic similarity of biotas and phylogenetic systematics. My goal in this final chapter is to reflect on the evidence and methods used to infer the temporal and spatial development of geographic distribution, with particular reference to the Caribbean.

Fossils

The fossil record has been thought to be to the biogeographer what DNA sequences are to the evolutionist. In principle, perfect and *complete* knowledge would fully describe the history and evolution of geographic distributions. Of course, this presupposes that the taxonomic and evolutionary relationships of all fossil and extant organisms could be correctly resolved. Recent studies on the Pleistocene and Holocene distribution of trees in North America and Europe, based on fossil pollen, are the best examples employing a relatively complete fossil record to reconstruct the historical dynamics of geographic distribution (Huntley and Birks 1983, Gandreau and Webb 1985, Webb 1987). However, for most organisms, at most times, and in most places, we are cursed with a very poor fossil record. Fossil distributions always underestimate the temporal and spatial occurrence of taxa (Raup 1979), and seldom preserve all the characters necessary to resolve their evolutionary relationships. What can fossils tell us about the history of geographic distributions?

Fossils contain information concerning character states and may serve in conjunction with neontological data as aids in reconstructing phylogeny. They are, however, no more valuable than neontological data for extant groups, and may be less so, if critical characters are not preserved (Patterson 1981). For extinct taxa, fossils obviously comprise the only source of information.

As windows on the past, fossils are most useful in documenting extinction.

Although they always underestimate the recency of extinction, their presence in regions where a taxon is now extinct can certainly alter our view of its biogeographic dynamics.

When their evolutionary relationships can be resolved, fossils can provide a crude estimate of the time during which taxa occupied a particular region. Yet because they provide only minimum estimates of the time when a group originated or went extinct, they can seldom controvert a biogeographic hypothesis. For example, if a taxon is proposed to have occupied the Greater Antilles since the late Cretaceous, a lack of fossils until the Eocene would be uninformative. This could be attributed to the notoriously poor quality of the fossil record rather than to the absence of the taxon at that time. However, if fossils were found at times earlier than the Cretaceous, then a hypothesis of occupation since the late Cretaceous would be refuted. In general, fossils can conflict with a biogeographic hypothesis only if they are found at times earlier than expected or in regions where they were not expected. Nevertheless, fossils still serve as the only direct means of linking a specific hypothesis concerning the dynamics of distribution to particular areas and particular times in the geologic past.

Phenetics

The assumption of phenetics is that similarity in appearance indicates close relationship. In the use of phenetic methods in phylogenetic inference, similarity in appearance is assumed to imply consanguinity. In biogeography, phenetic analyses of the similarities of biotas have been used to assess the degree of relationship between biotas at two or more geographic locations.

Phenetic similarity analyses involve preparing lists of the taxa present at each site within a region. The relationship between sites is estimated by either visual inspection of the lists or by the computation of some ad hoc or probablilistic statistic based on the number of taxa shared between sites (reviews in Connor and Simberloff 1978, Simberloff and Connor 1980, Simberloff et al. 1981, McCoy and Heck 1987). The higher the resemblance (the greater the total number of taxa shared), the more closely related two sites are considered. This procedure has been termed Q-mode analysis of biogeographic distributions (Simberloff and Connor 1980, Simberloff et al. 1981).

Biogeographic distributions can be represented as a two-dimensional binary matrix of presence and absence of taxa at sites. Sites correspond to columns and taxa to rows of the matrix. Q-mode analysis consists of pairwise comparison of columns (sites). For example, Figure 11-1 illustrates the distribution of 19 hypothetical genera among four sites. Table 11-1 shows

SITE

LINEAGE	A B C D	TOTALS
1	1 1 1 1	4
2	1 1 1 1	4
3	1 1 1 1	4
4	0 0 1 0	1
5	0 0 1 0	1
6	0 0 1 0	1
7	0 0 1 0	1
8	1 0 1 1	3
9	1 0 1 1	3
10	1 0 0 1	2
11	1 1 0 1	3
12	0 1 0 0	1
13	0 1 0 0	1
14	0 1 0 0	1
15	1 0 0 0	1
16	1 0 0 0	1
17	0 1 0 1	2
18	0 1 0 1	2
19	0 1 0 1	2
TOTALS	9 10 9 10	

Figure 11-1. Binary presence-absence matrix for 19 hypothetical genera distributed among four sites. Genera are numbered sequentially. Sites are designated by capital letters. 0's indicate absence and 1's indicate presence at a site.

the values of the simple matching coefficient (Sneath and Sokal 1973:132) when used to estimate the pairwise similarity between sites. A dendrogram showing the overall relationship among the four sites, constructed using the unweighted pair-group arithmetic averaging algorithm (Sneath and Sokal 1973:230), is pictured in Figure 11-2.

In ecological biogeography, site resemblance has been interpreted to indicate similarity in environments, and community membership. In historical biogeography, site resemblance has been assumed to indicate membership in the same biotic province (cf. Hagmeier and Stults 1964) and more importantly, a historical relationship between the biotas of the sites (e.g., Nichols, chap. 5, this volume). By "historical relationship" I mean the degree to which

Table 11-1. Values of simple matching coefficient from pairwise similarity analysis of data in Figure 11-1

		Sites			
		A	B	C	D
Sites	A	—			
	B	.42	—		
	C	.58	.32	—	
	D	.74	.68	.53	—

SITES

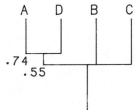

Figure 11-2. Phenogram generated using the simple matching coefficient and the unweighted pairgroup arithmetic averaging algorithm. The data used to generate this phenogram are depicted in Figure 11-1, and the values of the simple matching coefficient are presented in Table 11-1. The values at each node on the phenogram indicate the average level of similarity of all branches beyond that node.

the biotas of usually disjunct sites share a common evolutionary history. Yet, should this resemblance be construed to imply a historical relationship?

Phenetic similarity indexes may be poor estimates of historical relationship. This is because these indexes depend on differences between the sizes of the biotas at each site, and because they are very sensitive to differences between sites in the number of taxa found at only that site (single-site endemics). The diversity of all taxa at a particular site is controlled largely by forces operating on an ecological time scale (Hamilton and Rubinoff 1963, Hamilton et al. 1963, Strong et al. 1977). When historical relationships are estimated by phenetic similarity indexes, these forces combine to submerge the common evolutionary history of biotas among differences generated by recent extinctions, dispersal, and adaptive radiation. Savage (1982) has provided a numerical example of the effects of variation in the number of single-site endemics on estimated historical relationships between sites, and Connor and Simberloff (1978) discuss the dependence of similarity indexes on differences between sites in the diversity of taxa. These problems make phenetic similarity analysis of Q-mode biogeographic data unreliable as a tool for estimating historical relationships between sites.

Another kind of analysis of biogeographic data consists of pairwise, triowise, quartetwise, and so forth, comparisons of the distributions of taxa among sites. This is equivalent to a comparison of groups of rows (R-mode analysis) in the binary matrix described above (Connor and Simberloff 1979, 1983, 1984; Simberloff and Connor 1980). In ecological biogeography, R-mode analyses have been performed to determine if species or higher taxa exhibit independent distributions, or if commonly observed distributions are consistent with a hypothesis of interaction among taxa (competition, predation, or mutual association).

In historical biogeography, the method developed by Croizat et al. (1974), termed "generalized tracks," is essentially an R-mode analysis. Its purpose is also to estimate historical relationships, usually between disjunct sites. This approach consists of mapping the geographic distributions of monophyletic taxa and determining, by visual inspection, if groups of two or more sites are occupied unusually frequently by these taxa. The sites occupied by the component taxa in a single monophyletic group comprise a track (cf. Croizat et al. 1974, Figure 2). If many monophyletic groups display the same track, then this group of sites is termed a "generalized track." The difference between the method of generalized tracks and phenetic similarity analysis is that the former is based on sharing related taxa, while the latter is based on shared taxa regardless of relationship. I treat this approach under the heading of "phenetics" only because no detailed analysis of the phylogenetic relationship among component taxa within monophyletic lineages is required. However, one could as easily consider this a phylogenetic approach since it requires that the taxa examined be monophyletic.

Underlying the concept of the generalized track is the fact that all component taxa in a monophyletic group are related by descent. Therefore, groups of sites that each contain component taxa from a monophyletic lineage, and do so for a large number of lineages, must also be related historically. However, no specific ordering of the relationship among sites within a generalized track is possible using this method. Platnick and Nelson (1978), Rosen (1978), and many other historical biogeographers have rejected the method of generalized tracks largely for this reason. Yet the inference of historical relationship drawn from generalized tracks is sound, but the method of recognizing tracks is ad hoc. Before one can conclude that a track is indeed "generalized," one needs first to know that the number of monophyletic taxa having distributions consistent with that track is unusually high. Exactly what constitutes "consistency" with a track and what constitutes an "unusually" high number has yet to be determined.

One approach to testing for the presence of generalized tracks might be to represent the distribution of a population of monophyletic taxa, defined a priori, in a binary matrix, as described above. For a specified number of sites (m), the number of possible tracks of size $m - 1$, $m - 2$, ... 2 can be counted, and each can be enumerated. For example, for a region consisting of 6 sites there are 56 possible tracks ($2^m - m - 2$); 6 5-site tracks (6!(6 − 5)!5!), 15 4-site tracks, 20 3-site tracks, and 15 2-site tracks. There are many possible ways to assess the consistency of an individual monophyletic group's distribution with a particular track. One convention would be to consider a distribution consistent with a track if either the sites occupied by the monophyletic group are a subset of the track or if the track is a subset of the distribution of the monophyletic group. One could then count how many rows in the matrix (monophyletic groups) are consistent with each of the 56 possible tracks. The expected level of consistency with each track and its variance could be generated by tallying the same counts from randomly constructed binary matrices with row (and possibly column) sums identical to that of the actual data. The observed frequency of monophyletic groups

SITE

LINEAGE	A	B	C	D	TOTALS
1	1	0	1	1	3
2	1	0	1	1	3
3	1	0	0	1	2
4	1	1	0	1	3
5	0	1	0	1	2
6	0	1	0	1	2
7	0	1	0	1	2
TOTALS	4	4	2	7	

Figure 11-3. Reduced data matrix derived from Figure 11-1 by deleting uninformative taxa; those that occur at all sites, or single-site endemics.

consistent with each track could then be compared to the expected values to determine which, if any, track is in fact "generalized." This procedure has never been performed but could yield interesting and valuable results. The data necessary to perform such an analysis are readily available, and an efficient and unbiased algorithm for generating random binary matrices will be available soon (Simberloff and Zaman, in prep.). Alternatively, the probability of "missing" a particular number of tracks in the data at hand could be computed using Maxwell-Boltzmann statistics (Feller 1950, Simberloff and Connor 1981), but this approach lacks statistical power.

Using the data presented in Figure 11-1, one could ask whether certain tracks are underrepresented, or whether one would expect to observe the degree of consistency manifested by the data with each of the possible tracks. When uninformative lineages are removed from the matrix (single-site endemics and cosmopolitan taxa), the reduced data matrix pictured in Figure 11-3 is obtained. For regions with four sites there are ten possible tracks. If one uses the convention mentioned above, then five of the six two-site tracks are observed and only one is missing from the data (track BC). For three site tracks, two of the four possible are missing from the data (tracks ABC and BCD). Given seven taxa found at two or more sites, and three taxa found at three or more sites, the Maxwell-Boltzmann probabilities of missing these numbers of tracks, or a greater number, are 0.946 and 0.625, respectively. Had one adopted the convention that a distribution must be exactly congruent with a track to be consistent, then only the two-site tracks BD and AD are observed among the four taxa occupying two sites. The probability of this event, or one more extreme, is 0.167. The analysis would be unchanged for taxa occupying three sites.

A computer simulation based on the reduced data matrix with fixed row sums was also performed. One thousand randomly constructed matrices were generated and the number of taxa displaying distributions consistent with each of the ten possible tracks was tallied. Figure 11-4 illustrates the expected number of taxa consistent with each track (plus or minus two

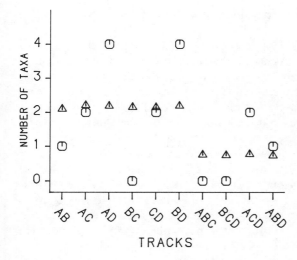

Figure 11-4. Comparison of observed and expected levels of consistency of distributions presented in Figure 11-3 with 10 possible generalized tracks. This is an example of an R-mode analysis for generalized tracks. The expected values (triangles) and their standard errors (bars) were generated from 1000 random binary matrices constructed subject only to the constraint that row sums equal those in the observed data. The observed level of consistency is depicted by circles.

standard errors), and the observed number of consistent taxa. If a hypothesis test with individual Type I error rates of $\alpha = 0.01$ is performed for each track, the experimentwide error rate is no greater than 0.096. In fact, because tracks such as AD and ACD are not independent, the true experimentwide error rate will be much lower than this nominal upper bound. Based on these analyses, one could conclude that tracks AD, BD, and ACD are overrepresented in the data and therefore may constitute generalized tracks. Tracks that are underrepresented in the data may define barriers to dispersal or regions whose biotas have largely independent evolutionary histories.

Phylogenetics

Rosen (1978) proposed an interesting and radically different approach to the analysis of historical biogeographic dynamics. Rosen's (1978) method relies mostly on modern techniques of phylogenetic inference, mainly cladistic analysis of shared, derived (mostly morphological) characters, rather than on an analysis of geographic distributions. The technique is again an R-mode analysis of monophyletic groups, but requires that a resolved phylogeny be constructed for each monophyletic lineage. If cladograms (phylogenies) for multiple lineages show the same pattern of ancestor-descendant relationships, and the most derived, next most derived, and so forth taxa from each lineage are found at the same sites, then Rosen (1978) concludes that these lineages display concordant patterns of evolution and biogeography. If this is observed, then the common cladogram describing the phylogeny of these lineages also depicts the order of relatedness among sites. Rosen (1978) also proposed a method of calculating the probability of concordance among clades of various sizes.

This method holds considerable promise for elucidating the historical relationships between sites. Although Rosen's (1978) protocol for calculating the probability of concordance is incorrect unless the clades to be analyzed are specified in advance, methods for estimating these probabilities have been developed (Simberloff et al. 1981, Savage 1983). However, because this approach requires that resolved cladograms be constructed for a very large number of clades, this technique has yet to be applied to the analysis of the historical biogeography of any region.

Another application of cladistic methods to the analysis of historical biogeographic relationships might be to perform a Q-mode cladistic analysis of biogeographic data. This would be analogous in form to the Q-mode phenetic analysis of binary presence-absence matrices described earlier, but would generate a cladogram of sites rather than a phenogram. This approach would use the presence or absence of monophyletic groups (say, all the species in a genus or all the genera in a family) at sites as character states. Lineages with taxa at only one site (single-site endemics) and those with taxa at all sites analyzed would be uninformative. Lineages that display restricted distributions, occupying between $m - 1$ and 2 of the m possible sites, would constitute the equivalent of derived characters. Sites that share many of these lineages with restricted distributions would by this technique be considered closely related (sharing derived characters). A cladogram of areas could then be generated using techniques outlined by Swofford (1984) or Felsenstein (1982). This method has been used to reconstruct the recent history of dispersal by species of freshwater fish in the Quebec peninsula by Legendre (1986). Brooks (1981) has also proposed a similar approach in employing parasite taxa as characters in constructing host cladograms.

The logic that underlies this approach is similar to that outlined for the method of generalized tracks. Sites that share related taxa are themselves considered related. The difference between this procedure and the method of generalized tracks is that, besides determining which sites are related, it also provides a hypothesis of the order of the relationship. Those sites that share the greatest number of monophyletic taxa, having restricted distributions, are considered to be most closely related. Taxa—both recently evolved and old—can have restricted distributions for two reasons: either because of a lack of expansion once evolved, or because of extinction in part of the geographic range. As a result, this procedure provides an estimate of the relationship among areas that includes historical as well as recent ecological effects on the distributions of monophyletic taxa. If vicariance and dispersal account for the distribution of most monophyletic taxa having restricted distributions, then this procedure will estimate the historical connections among areas. If restricted distributions are caused largely by recent extinc-

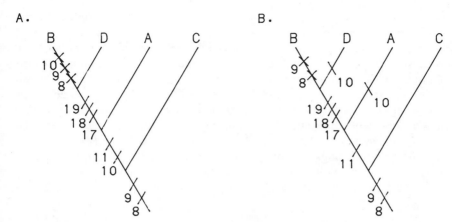

Figure 11-5. Two equally parsimonious area cladograms for hypothetical data presented in Figure 11-3. Using the notation of Sneath and Sokal (1973:333), dashes on the main stem indicate the presence of a genus at the site labeled on the subsequent branch, dashes on a branch indicate presence at that site, X's correspond to absences after presence (extinctions), and numbers correspond to the lineage numbers of Figure 11-3. A. Area cladogram in which lineage 10 is interpreted to be extinct. B. Area cladogram in which lineage 10 is interpreted to have dispersed recently to sites D and A.

tions, then this procedure will estimate the similarity in extinction history among sites.

The cladogram of sites depicted in Figure 11-5 was generated using the reduced data matrix of Figure 11-3 and the parsimony criterion (Felsenstein 1982). The cladogram is quite different than the phenogram estimated from the full-data matrix (Fig. 11-2), but suggests the same relationships between areas as the generalized track analysis presented above. This difference arises both because of the exclusion of widespread and endemic taxa and because of the parsimony criterion used in cladogram construction. Sites A and B switch positions between the cladogram (Fig. 11-5) and the phenogram (Fig. 11-2). In the phenetic analysis, the influence of single-site endemics dominates the structure of the phenogram, but in the cladogram single-site endemics play no role at all. The order of relationships depicted in the cladogram is based on the assumption that sites that share lineages with restricted distributions are most closely related.

If this procedure is applied to a number of higher taxa, say insect orders or families, then the set of cladograms generated from these analyses can be examined for their degree of concordance. An expected degree of concordance could be generated for randomly generated binary matrices with the same number of rows and columns, as well as row and, possibly, column sums as each of the higher taxa. The observed and expected levels of concordance could be compared to determine if a general pattern of relationships among sites is apparent.

This approach has never been previously suggested or applied to the problem of estimating historical biogeographic relationships. It provides a stronger basis of inferring historical relationships than the R-mode phenetic analysis of generalized tracks because it estimates the order of relationships among sites. However, it is more sensitive to recent dispersal and extinction than the analysis proposed by Rosen (1978). The greatest advantage of this procedure over R-mode cladistic analysis is that data are readily available and resolved phylogenetic cladograms need not be constructed for each lineage. Furthermore, R-mode cladistic analyses are based only on taxa that are endemic to a single site, while Q-mode cladistic biogeographic analysis ignores single-site endemics. Because of the independence of the data used in these analyses, they can be used in tandem with one approach serving to test the other.

Promise and Practice of Historical Biogeography

Of the four possible phenetic and cladistic Q- and R-mode biogeographic analyses described here (Fig. 11-6), three hold considerable promise for inferring historical biogeographic relationships between suites of geographic regions. Only phenetic Q-mode analysis is particularly unreliable. Of the remaining three, two could be applied to data that are currently available: R-mode phenetic analysis (generalized tracks) and Q-mode cladistic analysis (biogeographic inference of area cladograms). The cladistic character inference of area cladograms (Rosen's 1978 method) requires a data base not yet available for any geographic arena. Nevertheless, none of these techniques has been applied to the analysis of historical biogeographic relationships. What then do biogeographers do to infer historical relationships among sites?

As illustrated by the eight entomological contributions to this volume, biogeographers use a variety of approaches. Grimaldi, Hamilton, and Liebherr each performed a single cladistic character analysis of a particular family or genus of insects. This immediately precludes any probabilistic inference concerning concordance between clades. Since the taxa studied by these authors were not chosen with the goal of assessing their cladistic and biogeographic concordance, it would be improper to apply Rosen's (1978) method of computing probabilities after the fact. Even if these three taxa were exactly concordant in cladistic and biogeographic patterns, this could be a chance occurrence. To find two genera and part of one family of insects to display such patterns may not be at all unusual when sampling from the thousands of possible insect genera and families found in the Caribbean. Applying the protocol of Simberloff et al. (1981) would also be problematic. Given the small number of clades examined, any statistical test would proba-

Q-mode   R-mode

	Q-mode	R-mode
Ecological	Phenetic similarity analysis (Connor and Simberloff 1978)	Species co-occurrence analysis (Connor and Simberloff 1979)
Historical	A. Phenetic similarity analysis (Nichols, this volume) B. Cladistic biogeographic inference of area cladograms (this chapter)	A. Generalized track analysis (Croizat et al 1974 and this chapter) B. Cladistic character inference of area cladograms (Rosen 1978)

Figure 11-6. Types of Q-mode and R-mode analyses used in ecological and historical bio-geography. Note that phenetic similarity analyses have been used by both historical and ecological biogeographers. Cladistic biogeographic inference of area cladograms refers to the use of biogeographic data as "characters" in parsimonious construction of area cladograms. Cladistic character and biogeographic analysis involves using morphological or genetic data in parsimonious phylogenetic tree construction and subsequent concordance analysis to infer area cladograms.

bly lack the power to distinguish an unusually high degree of cladistic and biogeographic concordance from independent cladistic and biogeographic patterns. Of course, each of these cladistic analyses represents a considerable effort and is a valuable contribution to our knowledge of the evolution of insects in the Caribbean.

Once the cladogram was obtained, these authors resorted to a narrative discussion of the cladogram in relation to knowledge about other taxa, fossil data, ad hoc generalized tracks proposed by other authors, and to various scenarios concerning the geologic history of the Caribbean region. This process is actually quite similar in practice to that employed by Rosen (1978, 1985) and outlined by Platnick and Nelson (1978), since none of these authors actually performed statistical tests of the proposed relationships between areas. This, I believe, is the modus operandi of most historical biogeographers who employ R-mode cladistic analyses. In the absence of a sufficient population of resolved phylogenetic cladograms, as for the Caribbean, those few cladograms that have been constructed are bolstered by a hypothesis of relationships between areas that is based on the prevailing view of the geologic history of the region. As Donnelly's (see chap. 2) discussion of the history of the geologic history of the Caribbean illustrates, this is an ill-conceived and dangerous method for inferring historical biogeographic relationships. Geologic hypotheses are every bit as subject to change as biogeographic hypotheses. In essence, geologists have not and may not be able to resolve the area cladogram for the Caribbean. Until biogeographers can perform some sort of probabilistic analysis of purely biogeographic data, comparing single or small groups of cladograms to geologic scenarios is itself merely another scenario.

Of the other five entomological contributions to this volume, Nichols performs, interprets, and accepts the relationships among sites revealed by a phenetic Q-mode analysis. On the other hand, Slater performs yet casts aside such an analysis because of the disproportionate influence of single-site endemics. Slater and Eickwort also argue for relationships between areas based on historical examples of dispersal. Examples of recent dispersal are used to indicate the possibility of dispersal in the past. Ramos focuses on the high level of endemism exhibited by Homoptera in the Greater Antilles. Finally, Wilson emphasizes the role of extinction in shaping the modern ant fauna of the Caribbean as indicated by fossil evidence. However, when making statements concerning the origin or affinities of the Caribbean fauna, each of these authors rests his inferences on the distributions of related taxa. For example, Ramos infers a relationship between the Greater Antilles and Central America and Mexico because the closest relatives of the Antillean auchenorrhynchous Homoptera are found there. Slater infers relationships between the Antilles and West Africa because the closest relatives of the Antillean genus *Pachygrontha* are West African. In essence, these inferences constitute the proposition of distributional tracks (*sensu* Croizat et al. 1974), and differ from the analyses performed by Liebherr, Hamilton, and Grimaldi only in the absence of a cladistic analysis of the relationship among taxa. No effort was made to determine if the inferred tracks are indeed "generalized." This is equivalent to the problem mentioned above in the use of cladistic R-mode analyses. The track, like a simple cladogram, is merely one datum. Neither provides a firm basis for inferring the general relationships among areas. Without a substantial sample of either tracks or cladograms and without application of the statistical tests outlined above and by Simberloff et al. (1981), historical biogeographic inferences will continue to be narrative scenarios.

The procedures outlined here and by Rosen (1978) and Simberloff et al. (1981) provide a means to carry historical biogeography beyond scenario building. These techniques can be used to determine if monophyletic taxa display unusually concordant patterns of geographic distribution, thereby implying historical relationships between areas. What is needed to implement these procedures is a systematic data base containing the distributions, and preferably the cladograms, of many monophyletic taxa from a region like the Caribbean.

I do not mean to suggest that implementing and interpreting the results of these procedures will be a simple matter. Much biogeographic data remains to be collected and many taxa must be examined to determine if they are monophyletic. Even if unusually concordant patterns are revealed, the roles of dispersal, vicariance, and extinction in generating these patterns will continue to be debated. Furthermore, if Rosen's (1985) and Donnelly's (see chap. 2) suggestion that areas may have reticulate geologic histories is true, then it will be all the more difficult to discern concordant biogeographic patterns.

In sum, a variety of methods that permit probabilistic inferences concerning historical relationships among areas are available but largely unused by biogeographers. Instead, biogeographers are all too willing to give to the geologists the task of constructing area cladograms. However, it is best to examine the biogeography and evolution of the Caribbean biota independently of geologic hypotheses concerning area relationships. The procedures currently available permit the development and testing of area cladograms without geologic data. It is also possible that these techniques may shed some light on the mechanisms that underlie biogeographic patterns.

Acknowledgments

I wish to thank Jim Liebherr for stimulating me to think about the problems of biogeographic distributions and furthering my understanding of phenetic and cladistic methods. This manuscript benefited from thoughtful reviews by K. L. Heck, J. Liebherr, D. Simberloff, and F. Vuilleumier.

References

Brooks, D.R. 1981. Hennig's parasitological method: a proposed solution. Syst. Zool. 30:229–249.

Connor, E.F., and D. Simberloff. 1978. Species number and compositional similarity of the Galapagos flora and avifauna. Ecol. Monogr. 48:219–248.

———. 1979. The assembly of species communities: Chance or competition? Ecology 60:1132–1140.

———. 1983. Interspecific competition and species co-occurrence patterns on islands: Null models and the evaluation of evidence. Oikos 41:455–465.

———. 1984. Neutral models of species' co-occurrence patterns. In D.R. Strong, D. Simberloff, L.G. Abele, and A. Thistle, eds., Ecological Communities: Conceptual Issues and the Evidence, pp. 316–331. Princeton University Press, Princeton, N.J.

Croizat, L., G. Nelson, and D. Rosen. 1974. Centers of origin and related concepts. Syst. Zool. 23:265–287.

Feller, W. 1950. An Introduction to Probability Theory and Its Applications, vol. 1, 2d ed. John Wiley and Sons, New York.

Felsenstein, J. 1982. Numerical methods for inferring evolutionary trees. Quart. Rev. Biol. 57:379–404.

Gandreau, D.C., and T. Webb III. 1985. Late-Quaternary pollen stratigraphy and isochrone maps for the northeastern United States. In V.M. Bryant and R.G. Holloway, eds., Pollen Records of Late-Quaternary North American Sediments, pp. 247–280. American Association of Stratigraphic Palynologists, Dallas.

Hagmeier, E.M., and C.D. Stults. 1964. A numerical analysis of the distributional patterns of North American mammals. Syst. Zool. 13:125–155.

Hamilton, T.H., and I. Rubinoff. 1963. Isolation, endemism, and multiplication of species in the Darwin Finches. Evolution 17:388–403.

Hamilton, T.H., I. Rubinoff, R.H. Barth, and G.L. Bush. 1963. Species abundance: Natural regulation of insular variation. Science 142:1575–1577.

Huntley, B., and H.J.B. Birks. 1983. An atlas of past and present pollen maps for Europe: 0–13000 years ago. Cambridge University Press, Cambridge, Eng.

Legendre, P. 1986. Reconstructing biogeographic history using phylogenetic tree analysis of community structure. Syst. Zool. 35:68–80.

McCoy, E.D., and K.L. Heck. 1987. Some observations on the use of taxonomic similarity in large-scale biogeography. J. Biogeogr. 14:79–87.

Patterson, C. 1981. Significance of fossils in determining evolutionary relationships. Ann. Rev. Ecol. and Syst. 12:195–223.

Platnick, N.I., and G. Nelson. 1978. A method of analysis for historical biogeography. Syst. Zool. 27:1–16.

Raup, D.M. 1979. Biases in the fossil record of species and genera. Bull. Carnegie Mus. Natural Hist. 13:85–91.

Rosen. D.E. 1978. Vicariant patterns and historical explanation in biogeography. Syst. Zool. 27:159–188.

Rosen, D.E. 1985. Geologic hierarchies and biogeographic congruence in the Caribbean. Ann. Missouri Bot. Garden 72:636–659.

Savage, H.M. 1983. The shape of evolution: Systematic tree topology. Biol. J. Linnean Soc. 20:225–244.

Savage, J.M. 1982. The enigma of the Central American herpetofauna: Dispersals or vicariance? Ann. Missouri Bot. Garden 69:464–547.

Simberloff, D., and E.F. Connor. 1980. Q-mode and R-mode analyses of island biogeographic distributions: Null hypotheses based on random colonization. In G.P. Patil and M.L. Rosenzweig, eds., Contemporary Quantitative Ecology and Related Ecometrics, pp. 123–138. International Cooperative Publishing House, Fairland, Md.

———. 1981. Missing species combinations. Amer. Naturalist 118:215–239.

Simberloff, D., K.L. Heck, E.D. McCoy, and E.F. Connor. 1981. There have been no statistical tests of cladistic biogeographic hypotheses! In G. Nelson and D.F. Rosen, eds., Vicariance Biogeogrpahy: A Critique, pp. 40–63. Columbia University Press, New York.

Sneath, P.H.A., and R.R. Sokal. 1973. Numerical Taxonomy. W.H. Freeman and Company, San Francisco.

Strong, D.R., E.D. McCoy, and J.R. Rey. 1977. Time and the number of herbivore species: The pests of sugarcane. Ecology 58:167–175.

Swofford, D.L. 1984. P.A.U.P. Phylogenetic Analysis Using Parsimony, version 2.3. Illinois Natural History Survey, Champaign.

Webb, T. III. 1987. The appearance and disappearance of major vegetation assemblages: Long-term vegetational dynamics in Eastern North America. Vegetatio 69:177–187.

SUBJECT INDEX

Absent genera
Sweat bees, 240
Adaptive radiation, 254
Aggregation behavior, 194
Amber fossils, 40, 254–256
Ants, 222
Bees, 248
Chiapas, 184, 202
Dominican Republic, 184, 202, 223, 227, 248
Anachronism, 184
Anegada Trough, 29
Antennal structure
Fruit flies, 184–185
Antillean–Old World relation
Antilles–Asia, 188–189, 194
Arboreal species, 74, 124, 136, 221, 225
Arboricolous species
Ants, 225
Army ants
Dominican amber, 223, 227
Atlantic lysocline, 34
Autochthonous radiation, 11
See also Island radiation
Aves Ridge, 29, 34

Bahamas
Dispersal, 52
Scaritine carabids, 105
Baltic amber
Ants, 222
Fruit fly, 202
Beach species
Sweat bees, 241, 244
Belize, 26

Biogeographic exclusion, 74
Bonaire, 26
Brachyptery, 38, 43, 46, 72, 121, 136
Variation by island, 139
Bromeliad habitation, 145–146
Tarsal structure, 124, 136

Cactus feeding
Fruit flies, 190, 191, 200
Carbonate compensation depth, 23
Caribbean flood basalt, 22
Caribbean-Florida
Faunal relationship, 51
Caribbean Plate, 18, 19, 31, 33
Flood basalt, 22, 26
Cave habitation, 136
Cayman Trench, 8, 20, 21, 31, 33
Cenozoic
Left–lateral strike–slip, 26
Character analysis
Platynine carabids, 124, 147–149
Character assessment
Homoplasy, 123
Character displacement, 74
Character weighting
Platynine carabids, 132
Chiapan amber
Fruit fly, 202
Chortis Block, 17, 26, 27, 30, 33
Chromosomal characters
Fruit flies, 190
Cladistics, 6, 9, 53, 107, 123, 131, 163
Vicariance, 140–141
Cleptoparasitism
Sweat bees, 238, 239, 244, 246, 249

271

TAXONOMIC INDEX

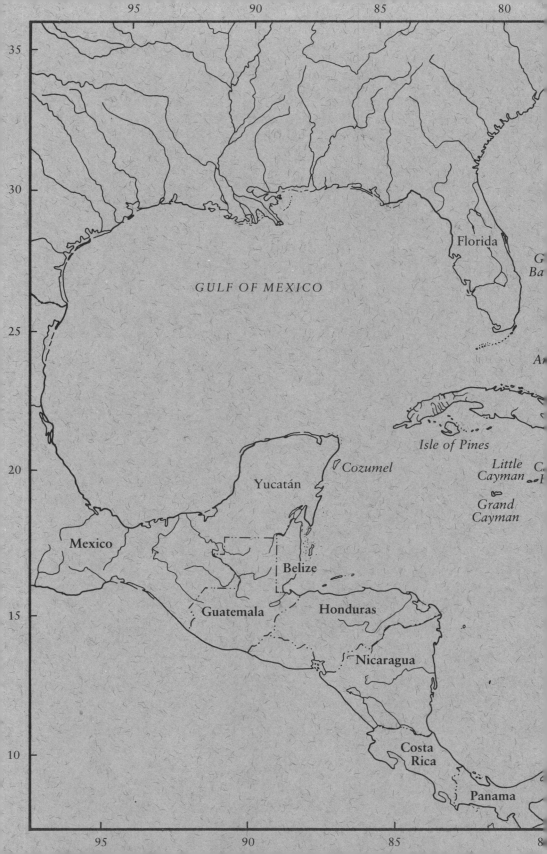